Road Development, Economic Growth, and Poverty Reduction in China

Shenggen Fan
and
Connie Chan-Kang

RESEARCH REPORT 138
INTERNATIONAL FOOD POLICY RESEARCH INSTITUTE
WASHINGTON, DC

International Food Policy Research Institute
2033 K Street, NW
Washington, DC 20006-1002 USA
Telephone +1-202-862-5600
www.ifpri.org

Library of Congress Cataloging-in-Publication Data

Fan, Shenggen.
Road development, economic growth, and poverty reduction in China / Shenggen Fan, Connie Chan-Kang.
p. cm. — (Research report ; 138)
Includes bibliographical references.
ISBN 0-89629-141-3 (alk. paper)
1. Roads—Economic aspects—China. 2. Infrastructure (Economics)—Government policy—China. 3. Rural poor—Government policy—China. 4. China—Economic policy—1976–2000. 5. China—Economic policy—2000– 6. China—Economic conditions—1976–2000—Regional disparities. 7. China—Economic conditions—2000—Regional disparities. I. Chan-Kang, Connie. II. International Food Policy Research Institute. III. Title. IV. Research report (International Food Policy Research Institute) ; 138.
HE365.C52F36 2005
338.951—dc22 2005006218

Contents

Tables

Foreword

In 1978, China initiated its economic and agricultural policy reforms. The ensuing rapid economic growth led to transportation shortages and congestion problems and increased the demand for roads. Since 1985 the government has given high priority to road development, particularly the construction of high-quality roads such as highways and freeways. While the construction of high-quality roads has taken place at a remarkably rapid pace, the construction of lower-quality and mostly rural roads has been slow.

This study evaluates the contribution of roads to economic growth and poverty reduction in China. It disaggregates road infrastructure into different classes of roads to account for quality, and then estimates the impact of road investments on overall economic growth, agricultural growth, urban growth, urban poverty reduction, and rural poverty reduction.

The study finds that benefit–cost ratios for lower-quality roads (mostly rural) are about four times larger than those for high-quality roads when the benefits are measured in terms of national GDP. Even in terms of urban GDP, these ratios are much greater for low-quality roads than for high-quality roads. In terms of poverty reduction, the study finds that, for every yuan invested, lower-quality roads raise far more rural and urban poor people above the poverty line than high-quality roads. Another significant finding of the study is the tradeoff between growth and poverty reduction in different parts of China, implying the need to formulate different regional priorities depending on whether economic growth or poverty reduction is more important for a particular part of the country.

IFPRI has long emphasized the importance of infrastructure for promoting economic growth and reducing poverty. Without this essential public good, efficient markets, adequate health care, a diversified rural economy, and sustainable economic growth will remain elusive. Effective development strategies require good infrastructure as their backbone.

The enormous benefit of rural roads that the study reveals for China holds true for other countries as well. Investment in rural roads should be a top priority to reduce poverty, maximize the positive effects of other pro-poor investments, and foster broadly distributed economic growth.

Joachim von Braun
Director General, IFPRI

Acknowledgments

This study is partially funded by the Asian Development Bank (ADB), the European Commission, the Australian Centre for International Agricultural Research (ACIAR), and the government of China through its support of the International Center for Agricultural and Rural Development (jointly sponsored by the Chinese Academy of Agricultural Sciences and the International Food Policy Research Institute). The authors have benefited tremendously from discussions with Eunkyung and Graham Gleave. Zhijun Zhao of the Chinese Academy of Social Sciences provided most of the highway length and cost data for this study.

The authors have benefited from discussion with Peter Hazell, who also provided guidance and numerous suggestions for improving the manuscript. IFPRI's Beijing office and later the International Center for Agricultural and Rural Development provided logistic support for field work and entered most of the data. Rowena Natividad provided professional assistance in formatting the report.

The IFPRI publication review committee coordinated the review of the manuscript. One external and one internal reviewer thoroughly assessed and reviewed the manuscript, for which the authors are grateful.

The views expressed in this report are those of the authors alone and should not be attributed to the ADB, the European Commission, ACIAR, or the government of China.

Summary

Beginning in 1978, China adopted a series of economic reforms leading to rapid economic growth and poverty reduction. National GDP grew at about 9 percent per annum from 1978 to 2002, while per capita income increased by 8 percent per annum. The post-reform period was also characterized by an unprecedented decline in poverty. However, income inequality has worsened between coastal and interior provinces as well as between rural and urban areas. A number of factors contributed to this widening disparity in regional development in China, including differences in natural resources endowments and infrastructure and human capital development.

When the policy reforms began in 1978, the transportation infrastructure in China was poor. With rapid economic growth, the demand for road transport soared, and transportation shortages and congestion problems surfaced as a consequence. Since 1985, the government has given high priority to road development, particularly construction of high-quality roads such as highways connecting major industrial centers in coastal areas. In the 1990s, investment in infrastructure became a national priority and various policies were implemented to promote the rapid construction of highways. The development of expressways has been particularly remarkable, with the total length increasing from 147 kilometers in 1988 to 25,130 kilometers in 2002, equivalent to an average annual growth rate of 44 percent. In contrast, the length of low-quality, mostly rural roads increased very little, by only 3 percent per year over the same period.

The objective of this study is to assess the impact of public infrastructure on growth and poverty reduction in China, paying particular attention to the contribution of roads. The beneficial impacts of roads on production and productivity, as well as on poverty alleviation, are well recognized in the literature but some important gaps remain. First, the impact of road quality has received little attention. While the total length or density of roads is a useful indicator of the road infrastructure availability in a country, it is important to account for quality differences because different types of roads (e.g., rural versus urban) can have very different economic returns and poverty impacts. Second, most studies have focused only on rural poverty in China, as urban poverty has only recently emerged as an important and growing problem. To address these limitations, this study disaggregates road infrastructure into different classes of roads to account for quality. The study also estimates the impact of road investments on overall economic growth, urban growth, and urban poverty reduction, in addition to agricultural growth and rural poverty. To achieve these goals, an econometric model that captures the different channels through which road investment impacts on growth and poverty is developed and estimated using provincial-level data for 1982–99.

The most significant finding of this study is that low-quality (mostly rural) roads have benefit–cost ratios for national GDP that are about four times greater than the benefit–cost ratios for high-quality roads. Even in terms of urban GDP, the benefit–cost ratios for low-quality roads are much greater than those for high-quality roads. As far as agricultural GDP is concerned, high-quality roads do not have a statistically significant impact while low-quality roads are not only significant but also generate 1.57 yuan of agricultural GDP for every yuan invested. Investment in low-quality roads also generates high returns in rural non-farm GDP. Every yuan invested in low-quality roads yields more than 5 yuan of rural non-farm GDP.

In terms of poverty reduction, low-quality roads raise far more rural and urban poor above the poverty line per yuan invested than do high-quality roads.

Another significant finding of the study is the trade-off between growth and poverty reduction when investing in different parts of China. Road investments yield their highest economic returns in the eastern and central regions of China while their contributions to poverty reduction are greatest in western China (especially the southwest region). This implies the need to formulate different regional priorities depending on whether economic growth or poverty reduction is the most important goal for the country.

The results of this study have important implications for future road project investments. In the past, China invested heavily in building expressways and intercity highways. These investments have been a major force in China's economic transformation during the 1980s and 1990s. However, as more and more investments are being poured into these projects, the marginal returns are beginning to decline, although they are still positive and economically sound. At the same time, low-quality roads or rural roads have received less attention than high-quality roads, and as a result, their marginal returns are much larger today than the returns for high-quality roads. Low-quality roads also raise more poor people out of poverty per yuan invested than high-quality roads, making them a win–win strategy for growth and poverty alleviation. The government should now consider giving greater priority to low-quality and rural roads in its future infrastructure investment strategy.

CHAPTER 1

Introduction

Since the mid-1980s, China has been investing massively in road infrastructure.[1] The resulting expansion of the road network, in addition to policy reforms and improvements in human capital, has been identified as one of the major engines of China's economic growth over the past decade (Fan, Zhang, and Zhang 2002). From 1985 to 2002, national gross domestic product (GDP) grew by more than 9 percent per year, making China's economy one of the most dynamic in the world.[2] While the Asian financial crisis unexpectedly hit the star performing economies of South Korea, Hong Kong, Singapore, Thailand, and Indonesia, the Chinese economy continued its growth momentum, with an annual GDP growth rate of 7–9 percent from 1998 to 2002.

Accompanying the rapid growth of the national economy was an astounding reduction of absolute poverty in rural China. Based on the official poverty line of $0.66 per capita per day (constant 1985 purchasing power parity [PPP] dollars), the incidence of poverty in rural China declined from 250 million people in 1978 to 29 million in 2001.[3] A reduction in poverty of this scale and within such a short time is unprecedented in history and is seen by many to be one of the greatest achievements in human development in the 20th century.

China still faces many challenges today, however. Prominent among these is that many people still live close to the official poverty line. For example, if, instead of using the official poverty line, poverty is measured using the international poverty line of US$1.00 per day (constant 1985 PPP dollars), then China still had more than 100 million rural poor and 20 million urban poor in 1998 (World Bank 2001). Another challenge is to implement policies that distribute economic gains more equally among the general population. Over the last two decades, the Gini coefficient index in rural areas increased from 0.21 in 1978 to 0.364 in 2002 according to data published by the State Statistical Bureau (SSB). For the country as a whole,

---

[1]Throughout this document, China refers to mainland China. Taiwan, Hong Kong, and Macau are not included because of unavailability of data and different economic systems.

[2]The data on GDP, income, and income distribution used in this report, if not otherwise specified, are from the latest issues of *China Statistical Yearbook* published by the National Statistical Bureau (Beijing: China Statistical Press).

[3]The number of rural poor for each year is reported in the *China Agricultural Development Report,* a white paper of the Ministry of Agriculture. The poverty line is defined as the level below which income (and food production in rural areas) is below subsistence levels for food intake, shelter, and clothing. Using this criterion, the rural poverty line in 1985 is 206 yuan in nominal price per person and increased to 625 yuan per person in 2000. The 206 yuan per year poverty line is equivalent to US$0.66 dollar per day measured in 1985 purchasing parity (World Bank 2001). China has never officially published the urban poverty rate although scattered reports on some cities are available using the ad hoc poverty lines.

the World Bank (2004) reported a Gini coefficient of 0.447 for 2001, making China one of the more unequal societies in Asia.[4] Another feature of the Chinese economy is the unequal development among regions. China has long pursued a biased development policy, with the largest portion of public investment concentrated in the coastal regions and in urban areas. It is not surprising therefore that the difference in economic growth rates between the coastal and inland regions was as high as three percentage points during the past two decades, and that regional inequality for China as a whole increased significantly.[5] Moreover, as a result of the government's urban-biased policy, the income and productivity gaps between rural and urban areas have increased over time. In 2002, rural per capita income was only one third of the urban per capita income. Thus, the majority of the poor people in China still reside in rural areas.

A large body of literature exists on economic growth and poverty reduction in China. However, few studies have attempted to link these issues to public investment and infrastructure development. It is widely recognized that improvements in human and physical capital are fundamental for economic growth and poverty reduction. Fan et al. (2002) were the first to link investments in infrastructure to rural poverty reduction in China. The authors used roads, electricity consumption, and the number of rural telephones as proxy variables for rural infrastructure in an econometrically estimated equations system. Their analysis of road infrastructure was quite crude, however, as the road variable was expressed in terms of the total length of all types of roads and failed to discriminate between roads of different quality.

This study builds on the earlier work of Fan et al. (2002) and uses a similar econometric model fitted to time series data at the provincial level. However, the model used here provides a more disaggregated analysis of the impact of road investments on economic growth and poverty alleviation in China. Roads are disaggregated into different classes to account for differences in their quality, and poverty impacts are measured separately for the rural and urban populations.

The report is organized as follows. The next chapter reviews the trends in economic growth and poverty in China. The third chapter reviews the development of China's road infrastructure, encompassing a review of the government's policies concerning roads, trends in road investment and construction, and the means by which they were financed. The fourth chapter provides a literature survey of the impact of road investments on growth and poverty reduction. The fifth chapter then presents the conceptual framework and model used, followed by a description of the available data and the model results. We conclude with a summary of the key findings and a discussion of their policy implications.

[4]This coefficient is among the highest in Asia but is lower than that found in some countries in Latin America and Africa. For example, the reported Gini ratio was 0.325 for India in 2000, 0.249 for Japan in 1993, 0.318 for Bangladesh in 2000, 0.591 for Brazil in 1998, and 0.593 for South Africa in 1995 (World Bank 2004).

[5]Kanbur and Zhang (2004) used both the Gini and maximum entropy indexes in assessing the changes in regional inequality (among provinces). They found that both measures increased substantially in the 1980s and 1990s.

CHAPTER 2

Economic Growth, Regional Development, and Poverty Reduction

This chapter provides an overview of China's economic growth, regional development, and poverty trends in recent decades. It also highlights the institutional and policy environment in which these changes took place. The analysis draws on Chinese official documents including various issues of *China Development Report* (National Bureau of Statistics), *China Agricultural Development Report* (Ministry of Agriculture), *China Rural Poverty Monitoring Report* (National Bureau of Statistics), and *An Overview of the Development-Oriented Poverty Reduction Program for Rural China* (State Council Leading Group Office of Poverty Alleviation and Development).

Economic Growth

China's economy has experienced major structural changes and economic transformations since the establishment of the People's Republic in 1949. The history of China's economic development is typically divided into two broad periods: the central planning period from 1949 to 1978, and the market-oriented reforms period from 1978 to the present.

The Central Planning Period, 1949–78

Following the establishment of the People's Republic of China in 1949, the Communist Party became the ruling party, controlling the country's political life and socioeconomic activities. The leaders' goals were to transform China into a modern, powerful, and socialist nation. The party initially adopted the Soviet economic model of communism, with its emphasis on a heavy industrial base, state ownership of the means of production in urban areas, collectively owned large units in agriculture, and centralized economic planning. At first, industrial GDP grew at a remarkable 16.1 percent per annum between 1952 and 1957, whereas total GDP grew by 9 percent per annum and the agricultural sector by 5 percent per annum. The government gradually took control of the industrial and agricultural sectors of the economy and by the end of the 1950s, privately owned firms were practically nonexistent. In the agricultural sector, land was confiscated by the state and merged into large and socialized production cooperatives, which evolved into communes in 1958. The commune was based on the collective ownership of all land and major inputs by its members, who produced mainly to meet state planning targets.

In the late 1950s, the authorities decided that China should make a "great leap forward" and modernize. The goal was to depart from the Soviet model and catch up with the United Kingdom in 15 years. All the country's resources were mobilized to reach this goal. This unrealistic campaign resulted in a severe economic crisis, and GDP declined precipitously at a rate of 8.7 percent per annum between 1958 and 1962. Agricultural GDP also fell by 3.1 percent per annum and industrial GDP by 14.3 percent per annum. The Great Leap Forward campaign turned out to be a disaster when the country was hit by a series of natural calamities and tens of millions of people died in the ensuing famine. Between 1962 and 1965, the Chinese economy entered a period of readjustment and recovery, driven by a series of corrective measures. Particularly noteworthy was the decentralization of decision making within the commune structure. The production brigades and teams became accountable for all decisions concerning production and the distribution of income among their members. In the industrial sector, production decisions were based on rational and efficient planning rather than on ideology. These changes were beneficial to the Chinese economy, and GDP in all sectors grew to their levels prior to the Great Leap Forward.

These developments were suppressed again with the Cultural Revolution (1966–76) when China was run under a strict Maoist vision. The government restored most of the controls that had been relaxed during the three-year adjustment period of 1962–65 and all foreign ties were cut off, isolating China from the rest of the world. During this period, the pace of growth in all sectors of the economy slowed down.

In sum, by the end of the Cultural Revolution, China was a centrally planned economy. The central government not only controlled nearly all means of production, but also dictated and controlled the economy through a variety of regulatory mechanisms. The government set production goals, controlled prices, and allocated resources. As self-reliance was emphasized, foreign investment was discouraged and trade became negligible. Despite these constraints, during China's communist regime all sectors of the economy expanded at a respectable rate. Between 1952 and 1977, real GDP in the agricultural sector grew by 4.2 percent per annum, whereas the corresponding rates in the industrial and service sectors were 8.1 percent per annum and 4.0 percent per annum, respectively (Table 2.1).

The Reform Period, 1978–Present

In December 1978, China's party leaders decided to embark on a program of gradual but fundamental reform of the economic system. Underlying the reform was a principle of incrementalism or gradualism: new measures were first to be implemented locally, and if they proved successful, would be popularized and disseminated nationally.

The first stage of the economic reform program (1978–84) focused mainly on the rural sector. The government aimed at expanding agricultural production, diversifying the rural economy, improving the rural standard of living, and promoting new technologies. One of the most important and successful reform policies was the implementation of the household responsibility system, which allowed farmers to have use rights over land. Collectively owned and operated land was distributed to farmers based on family size and the number of family workers. Farmers had the right to choose their own crop mix and input levels as long as they fulfilled government quotas for certain grain and cash crops. Any surplus output could be sold in the free markets. This system motivated farmers to reduce production costs and to increase productivity, since their efforts were closely linked to their income. Accompanying increases in procurement prices and government encouragement of the establishment of free farmers' markets stimulated agricultural production further.[6]

[6]The forced procurement policy for grains and certain cash crops was implemented from the 1950s to 1978. Under this policy, production teams had to sell a certain amount of their outputs to the government (the so-called quota). The procurement (or quota) prices were usually much lower than the international prices and free market prices. In 1978, the government decided to increase the quota price by 20 percent and the price for over-quota grain by an additional 50 percent (*China Agricultural Development Report,* 1997).

Table 2.1 Gross domestic product by sector

	GDP total	Agriculture	Urban industry	Urban service
		(billion yuan, 2002 constant prices)		
1952	259	131	54	74
1957	403	162	120	121
1962	364	143	114	107
1965	555	211	195	150
1970	776	273	314	188
1975	1,033	335	473	226
1978	1,222	343	589	290
1979	1,315	410	623	282
1980	1,418	427	688	303
1981	1,491	474	692	326
1982	1,627	541	732	353
1983	1,804	596	805	404
1984	2,079	665	900	513
1985	2,359	669	1,018	673
1986	2,567	695	1,130	741
1987	2,865	767	1,258	840
1988	3,189	818	1,407	963
1989	3,319	830	1,429	1,061
1990	3,445	932	1,434	1,080
1991	3,762	920	1,584	1,258
1992	4,297	936	1,887	1,474
1993	4,877	969	2,313	1,594
1994	5,491	1,111	2,627	1,753
1995	6,068	1,244	2,961	1,862
1996	6,650	1,356	3,293	2,001
1997	7,235	1,381	3,617	2,238
1998	7,800	1,449	3,845	2,506
1999	8,357	1,474	4,130	2,753
2000	9,026	1,476	4,533	3,017
2001	9,703	1,537	4,861	3,306
2002	10,479	1,612	5,354	3,513
Annual growth rate (percent)				
1952–2002	7.62	5.68	8.94	7.90
1978–2002	9.76	6.33	10.25	11.93
1952–1977	5.43	4.24	8.08	4.01
1978–1983	7.83	11.14	6.04	7.09
1984–1991	8.59	5.53	8.02	12.45
1992–2002	9.02	5.43	10.16	9.41

Source: Authors' calculations based on China's National Bureau of Statistics, *Statistical Yearbook of China*, 2001, 2003.

Notes: Since there is no official GDP deflator, we calculated the real GDP by multiplying the real GDP index listed in the *Statistical Yearbook of China* by the 1952 base year value. To calculate the GDP by sector, we multiplied the real GDP by the respective shares of each sector in total nominal GDP.

In the industrial sector, policies were introduced to increase the autonomy of enterprise managers, to reduce the prominence of planned quotas, and to allow enterprises to produce and sell goods in the market. Township and village enterprises (TVEs), which are industries owned by townships and villages, were established and individual enterprise was allowed after having virtually disappeared during the Cultural Revolution.

Moreover, China introduced an open-door policy permitting international trade and foreign direct investment.

These initiatives improved the standard of living for most Chinese. Income as well as the availability of food, housing, and other consumer goods increased substantially. Between 1978 and 1983, real GDP grew on average by 7.8 percent per annum, and agricultural and industrial output grew at 11.1 percent and 6.0 percent, respectively (Table 2.1).

The second phase of the reform program (1984–91) was aimed at broadening the reforms to include industrial enterprises in urban areas, creating market institutions, and dismantling the central planning system. Two particularly important policies were introduced: the dual-track pricing system for industrial goods and the enterprise contract responsibility system. Under dual-track pricing, some goods and services were allocated at state controlled prices, while others were allocated at market prices. Prices were gradually deregulated while markets were allowed to play an increasing role in setting prices. The enterprise contract responsibility system granted greater autonomy to production and employment decisions within enterprises.

Another key element of the reforms was to allow private as well as foreign enterprises to compete with state-owned enterprises (SOEs). The development of the non-state sector in China provided employment opportunities and contributed to the country's economic growth. Other important measures included the dissolution of China's monobank system under which the People's Bank of China served as both the central bank and the sole commercial bank, the introduction of an enterprise tax system, and an expansion of the number of special economic zones. These policy changes proved beneficial to the industrial and services sectors, which grew rapidly between 1984 and 1991 by 8.0 percent and 12.5 percent per annum, respectively (Table 2.1).

The third phase of the reform program (1992 to the present) aimed at establishing a socialist market economic system, in which the economy continues to remain primarily under public ownership but market forces are allowed to play a fundamental role in resource allocation and distribution decisions.[7] It was the first time that China officially endorsed the market economy. To achieve this goal, several measures have been taken to reform the financial and fiscal sectors. In the financial sector, a key focus of the strategy was to create a banking system engaged in commercial transactions and responsive to market forces, shifting away from the banks' traditional role of supporting SOEs. In the fiscal sector, measures were directed to improve tax administration and set up stable and transparent tax rates. Since the mid-1990s, Chinese authorities have been focusing on improving the efficiency and profitability of the SOEs and on developing the social security system.

China's accession to the World Trade Organization (WTO) in late 2001 is an important step in the country's transformation into a market-driven economy. In accordance with its membership in the WTO, China will have to further open and liberalize its economy in the coming years. Trade barriers will have to be lowered, financial markets and institutions further developed, and the private sector further deregulated. China will also have to continue improving the performance of state-owned firms.

During the past 50 years, the economic structure of China has undergone tremendous changes. In 1952, China's economy was essentially based on agriculture. The agricultural sector accounted for half of

[7]The third phrase began when Deng Xiaopeng made an important speech during his South China tour. He reaffirmed that reform would have to continue, and it was the first time that the leader recognized the importance of the market economy in China's economic development instead of the socialist economic system.

Table 2.2 GDP and employment shares by sector

	Share of GDP by sector			Share of employment by sector		
	Agriculture	Urban industry	Urban service	Agriculture	Urban industry	Urban service
	(percent)					
1952	50.5	20.9	28.6	83.5	7.4	9.1
1957	40.3	29.7	30.1	81.2	9.0	9.8
1962	39.4	31.3	29.3	82.1	7.9	9.9
1965	37.9	35.1	27.0	81.6	8.4	10.0
1970	35.2	40.5	24.3	80.8	10.2	9.0
1975	32.4	45.7	21.9	77.2	13.5	9.3
1978	28.1	48.2	23.7	70.5	17.3	12.2
1979	31.2	47.4	21.4	69.8	17.6	12.6
1980	30.1	48.5	21.4	68.7	18.2	13.1
1981	31.8	46.4	21.8	68.1	18.3	13.6
1982	33.3	45.0	21.7	68.1	18.4	13.4
1983	33.0	44.6	22.4	67.1	18.7	14.2
1984	32.0	43.3	24.7	64.0	19.9	16.1
1985	28.4	43.1	28.5	62.4	20.8	16.8
1986	27.1	44.0	28.9	60.9	21.9	17.2
1987	26.8	43.9	29.3	60.0	22.2	17.8
1988	25.7	44.1	30.2	59.4	22.4	18.3
1989	25.0	43.0	32.0	60.0	21.6	18.3
1990	27.0	41.6	31.3	60.1	21.4	18.5
1991	24.5	42.1	33.4	59.7	21.4	18.9
1992	21.8	43.9	34.3	58.5	21.7	19.8
1993	19.9	47.4	32.7	56.4	22.4	21.2
1994	20.2	47.8	31.9	54.3	22.7	23.0
1995	20.5	48.8	30.7	52.2	23.0	24.8
1996	20.4	49.5	30.1	50.5	23.5	26.0
1997	19.1	50.0	30.9	49.9	23.7	26.4
1998	18.6	49.3	32.1	49.8	23.5	26.7
1999	17.6	49.4	32.9	50.1	23.0	26.9
2000	16.4	50.2	33.4	50.0	22.5	27.5
2001	15.8	50.1	34.1	50.0	22.3	27.7
2002	15.4	51.1	33.5	50.0	21.4	28.6

Source: Authors' calculation based on data reported by China's National Bureau of Statistics, *Statistical Yearbook of China,* 2001, 2003.

national GDP, whereas the industrial and service sectors contributed 21 percent and 29 percent, respectively (Table 2.2). By 2002, China had emerged as an industrial economy with 51 percent of GDP derived from industry, 34 percent from services, and only 15 percent from agriculture. Nevertheless, in 2002 the agricultural sector still employed 50 percent of the labor force and remains an important component of China's economy.

In sum, since the initiation of economic reforms, China's economy has grown at an impressive rate of 9.8 percent per annum. Most Chinese have benefited from rising income levels and living standards, and China is now classified as a lower-middle-income country. However, a number of challenges remain. In the financial sector, for example, inefficient state-owned banks still dominate the sector and continue to fund the SOEs. Although the viability of SOEs has been challenged with increasing competition from domestic and foreign firms, they continue to monopolize some sectors such as heavy industry and utilities. Moreover, the social security system is not yet functioning well. Problems include insufficient funding and limited and insufficient coverage of beneficiaries. In the agricultural sector, additional reforms are needed, as most farms remain very small and use low-level technology. Moreover, most farmers still do not have secure title to their land.

Regional Development

For a country as large and as geographically diverse as China, uneven regional development is far from unexpected. The coastal area, endowed with favorable geographical and natural conditions, has historically developed faster than the interior regions. Over the years, China's development policies have exacerbated this unequal development pattern. During the central planning period, the government pursued a strategy of regional economic self-sufficiency, which significantly shaped regional economic outcomes. Each region was expected to be self sufficient in terms of both food production and industrial goods. This policy severely distorted the allocation of resources, which in turn affected production efficiency and aggravated the unevenness in regional economic development. The introduction of economic reforms further shifted China's focus away from developing the interior provinces. To attract foreign direct investment and to promote foreign trade, the government established special economic zones along the coast. The resulting rapid economic growth in the coastal provinces continued to widen the gap between the coastal and the interior regions.

Differences in economic development among regions can be gauged by their GDP per capita, agriculture's share in regional GDP, and rural income per capita (Table 2.3). In 2002, per capita GDP in the northwest and southwest averaged 5,000–6,000 yuan, or only about half the income level in the east, the northeast, and the northern regions. The rural income per capita in northwest (1,744 yuan) and southwest China (1,894 yuan) was also considerably lower than the national average (2,476 yuan) and that in the east (3,203 yuan). Moreover, the shares of agriculture in total GDP are higher in the northwest and southwest (18 and 21 percent) than in the east and north regions (12 and 10 percent), indicating that farming is still a major source of rural income in western China.

A combination of economic, social, geographical, and others factors has restricted

Table 2.3 Major economic indicators by region, 2002

	Population	Per capita GDP	AgGDP/GDP	Per capita income: Urban	Per capita income: Rural
	(million)	(yuan)	(percent)	(yuan)	
North	124.59	10,758	10	8,296	2,703
Northeast	107.15	10,813	13	6,315	2,509
East	367.61	12,266	12	8,823	3,203
Central	308.92	9,018	15	8,505	2,641
Southwest	250.39	5,144	21	6,872	1,894
Northwest	116.52	6,180	18	6,304	1,744
China	1,275.18	9,255	14	7,703	2,476

Source: Authors' calculations based on data reported by China's National Bureau of Statistics, *Statistical Yearbook of China,* 2003.

Note: Rural and urban per capita income are weighted average figures that are obtained by multiplying the average per capita income by province reported in the 2003 yearbook by the region population shares for 2000.

labor mobility in China's less developed regions and this has been a major contributor to the widening regional disparities (Kanbur and Zhang 1999). Differences in the shares of the rural labor force employed by the rural non-farm sector are also seen as important (Rozelle 1994). For the nation as a whole, in 1997, about 29 percent of the rural labor force was engaged in non-agricultural activities and the non-farm sector contributed more than one third of rural income. But in the east, the non-farm sector employed 40 percent of the rural labor force, compared with less than 20 percent in northwest and southwest China.

Given that such an overwhelmingly large share of the rural labor force is employed in agricultural activity in China's western region, labor productivity is inevitably low. In fact, labor productivity in the southwest was half the national level in 1997. Poor natural resource endowments, weak infrastructure, low literacy rates, and insufficient investment and personnel in regionally focused science and technology research all constrain the development and adoption of new technologies and associated improvements in agricultural productivity. Moreover, difficulties in accessing national and international markets constrain the choice of cropping mix and the development of high-value agricultural products in western China (Fan et al. 2001a).

The growing inequality and increasing concentration of poverty in the western region led the central government in 1999 to launch an official plan aimed at developing western China. Among the specific objectives included in the proposed plan are the improvement of infrastructure; the intensification of environmental protection; and the development of science, technology, and education.

Overall, regional inequality in China can be lessened by adopting policies that promote growth in rural areas, that correct the bias in regional development, and that facilitate labor mobility across regions and from rural to urban areas. The latter is especially important to help narrow the wage gap between the farm and non-farm sectors, to transfer resources to rural areas (through remittances, investments, human capital, and information), and to improve the allocation of labor from low- to high-productivity sectors. The economic and social benefits of labor migration have been recognized by Chinese authorities. Recently, the government has freed up the movement of the labor force across China, opening cities to permanent migration.

Poverty

China has achieved tremendous success in reducing poverty since the introduction of economic reforms in the late 1970s. Based on China's official poverty line of US$0.66 per day (in constant 1985 purchasing power parity [PPP] dollars), the number of rural poor decreased dramatically from 250 million in 1978 to 29.3 million in 2001, implying that more than 200 million people escaped from poverty in the past 23 years (Table 2.4).[8] The reduction in rural poverty was particularly rapid during the first phase

[8]The accuracy of China's official poverty measure has been criticized in the literature for a number of reasons. For example, the difference in the cost of living among regions or provinces is not taken into account because a single poverty line is used for the whole of China. Moreover, the estimation of the official poverty line is based on income rather than on expenditure data and has not kept pace with inflation over the years. As a result of these limitations and other problems, the official estimates may be biased as they tend to overstate the decline in poverty and understate the remaining incidence (for more details see Riskin 2004). Nevertheless the literature overall agrees that the absolute rural poverty has declined greatly over the past several decades. Moreover, even using the higher poverty line of US$1.00 per day measured in purchasing power parity, the poverty rate declined dramatically from 31.3 percent in 1990 to 11.5 percent in 1998, or the number of poor declined from 280 million to 106 million (World Bank 2000).

Table 2.4 Rural poverty in China, 1978–2000

Year	Number of poor	Poverty incidence
	(million)	(percent)
1978	250	33.1
1984	128	15.1
1985	125	14.8
1992	80.1	8.8
1993	75	8.2
1994	70	7.6
1995	65	7.1
1996	58	6.3
1997	49.6	5.4
1998	42.1	4.6
1999	34.1	3.7
2000	32.1	3.4
2001	29.3	3.2

Source: China Statistics Press, *China Rural Poverty Monitoring Report* (various issues).

of the reform period when increases in agricultural production, productivity, and prices stimulated the rural economy. By the end of 1984, the year marking the end of the first phase of the reforms, the number of rural poor (128 million) was half the level of 1978 (250 million). Likewise, the rural poverty rate fell dramatically during this period, from 33.1 percent in 1978 to 15.1 percent in 1984 (Table 2.4).

During the second stage of the reform, the government introduced the first important poverty reduction program. The government allocated funds for public works with the goal of boosting income in poor counties and stimulating local economic growth. Rural poverty continued to decline rapidly during this period. By the end of the second phase of the reform period in 1992, rural poverty had declined to 80.1 million and the incidence of rural poverty had fallen to 8.8 percent (Table 2.4).

In the mid-1990s, China launched an ambitious poverty alleviation plan aimed at eradicating poverty by the end of the 20th century. Some key features of the program included subsidized loans, food-for-work infrastructure schemes, and various grants, which were targeted to 592 nationally designated poor counties.[9] The government directly supported agriculture by providing low-interest loans. Poverty declined rapidly during this period, from 70 million in 1994 to 29 million in 2001, while the incidence of rural poverty fell from 7.6 percent to a mere 3.2 percent (Table 2.4).

However, estimates of the extent of poverty reduction are sensitive to the choice of the poverty line. If poverty is measured using the World Bank's international poverty standard of US$1.00 a day (in 1985 purchasing power parity dollars) instead of the official Chinese poverty line, then, by that criterion, although poverty is still found to have declined rapidly since the beginning of the reform period, a much larger number of poor remain today (106 million rural poor in 1998, equivalent to a poverty rate of 11.5 percent) (World Bank 2001). Clearly, this much higher poverty figure demonstrates that a large number of rural people live just above the official poverty line, and thus pose an important and remaining challenge for the Chinese government.

When characterizing poverty in terms of nutrition, evidence shows that there was a remarkable improvement in child nutrition in rural China in the 1990s. Between 1990 and 1998, the incidence of underweight children younger than five years of age declined from 22 percent to 12.6 percent in rural China, while the incidence of stunting fell from 41.4 percent to 22 percent. These achievements were partly driven by the rapid socioeconomic developments of the 1990s including the various poverty alleviation and other special programs that were put in place.

---

[9]The detailed proposed poverty reduction programs were described in the book *An Overview of the Development-Oriented Poverty Reduction Program for Rural China* (State Council Leading Group Office of Poverty Alleviation and Development 2003).

Rural poverty is not evenly distributed in China, as shown in Table 2.5. Regardless of the sources of data, it is clear that the incidence of rural poverty is significantly higher in western China, a region characterized by poor agricultural land and weak infrastructure. Based on the World Bank's estimates, the poverty rate averaged 19 percent and 11 percent in the northwest and southwest regions, respectively, in 1996, while the comparable figure for the entire country was 6 percent. In contrast, for that same year, only 1.2 percent and 2.6 percent of the rural population was living under the poverty line in the eastern and central regions, respectively. Moreover, the reduction in the incidence of rural poverty was not as dramatic in western China as it was in the other regions. Consequently, rural poverty is increasingly concentrated in the western provinces. The southwest and northwest regions together accounted for about 70 percent of the total rural poor in China in 1996, a 40 percentage point increase over 1988 (Table 2.5).

Although urban poverty has recently become an important policy issue, its relatively low level in previous years was attributable to various urban-biased policies of the Chinese authorities since the 1950s. First, as part of the country's industrial development strategy, various implicit and explicit transfer programs favoring the urban sector were implemented (Lin, Cai, and Li 1996; Hussain 2003).[10] The urban sector benefited from the economic growth led by industrialization, at the expense of the countryside. Moreover, the rationing system introduced in the 1950s enabled urban residents to have equal access to food and other necessities at much lower prices than rural people. Almost all urban residents of working age also had guaranteed jobs in the state- or collective-owned sectors. Because these jobs were permanent ("iron rice bowl"), urban unemployment was virtually nonexistent. These jobs also provided urban residents with many benefits such as free or subsidized housing and health care. Not surprisingly, poverty alleviation in urban areas was not on the policy agenda until recently, and China's anti-poverty program, first initiated in 1986, focused mainly on rural areas.

China's recent economic reforms have been beneficial to urban dwellers in general but have also contributed to worsening poverty and inequality for some. One key feature of the reforms was the reduction of workers' lifetime ties to their employers, thereby providing them with a higher degree of freedom to change jobs and achieve higher incomes. The reforms also allowed market forces to determine wages and better match wages with workers' skills. The efficiency gains from the urban reform are evidenced by a dramatic increase in the average per capita urban income, which grew at about 6 percent per year in the 1990s. On the other hand, urban reforms and increased competition have resulted in soaring financial losses for many state- and collective-owned enterprises, and an increasing number of urban workers have been laid off.[11] Many of these workers are not adequately compensated by the existing social safety net programs. Also, the liberalization of the welfare system may have made some disadvantaged groups more vulnerable to sudden shocks such as catastrophic illness. Nowadays, the urban poor in China can be grouped into the following three categories: (1) urban people with no working ability, no income, or no providers; (2) unemployed workers; and (3) low-income or temporarily laid-off workers. To mitigate the surge of urban unemployment and poverty, the

[10]This bias still exists today, but in different forms (for example, the government spends more in urban than in rural areas, universities have higher admission scores for rural students, and there are still visible and invisible restrictions on rural–urban migration).

[11]The number of laid-off workers in 1997 is reported to be 11.57 million (*China Development Report* 1998).

Table 2.5 Regional distribution of rural poor

	Xian and Sheng (2001)	World Bank (2001)			
	Poverty incidence	Poverty incidence		Share of national total	
	1998	1988	1996	1988	1996
North		15.2	4.6	9.8	7.6
Beijing	2.37	0.5	0.8	0	0.1
Tianjin	16.16	1.8	0.3	0.1	0
Hebei	39.11	14	3.9	5.8	4.1
Shanxi	44	22.1	7.5	3.9	3.4
Northeast		10.2	4.7	4.6	5.1
Liaoning	13.04	8	2.9	1.5	1.3
Jilin	18.18	8.8	4.7	1.1	1.3
Heilongjian	22.84	13.6	6.7	2.1	2.5
East		7	1.2	14.9	6.4
Shanghai	0.45	0.3	0.1	0	0
Jiangsu	12.37	4.5	0.1	2	0.1
Zhejiang	2.5	2.8	0.1	0.8	0.1
Anhui	27.64	10.3	2.7	3.9	2.7
Fujian	2.73	3.9	0.5	0.8	0.3
Jiangxi	9	8.5	0.7	2.1	0.5
Shandong	15.58	9.4	1.9	5.3	2.7
Central		13.6	2.6	23.1	11.3
Henan	31.75	25	4.3	14.6	6.6
Hubei	16.33	11	2.7	3.6	2.1
Hunan	5.08	7.9	1.5	3.4	1.6
Guangdong	2.46	2.7	0.2	1.1	0.2
Hainan		12.1	8.2	0.4	0.8
Southwest		20.5	10.5	31.9	41.2
Chongqing			6.6		3.2
Sichuan	21.49	16.7	7	12.8	9.7
Guizhou	42	23	12.8	5.3	7.6
Yunnan	36.45	23.8	22.9	6.2	15.3
Tibet	57.33	32.3	10.1	0.5	0.4
Guangxi	20.08	24.1	6.4	7.2	5
Northwest		26.2	18.6	15.7	28.4
Inner Mongolia	19.78	17.3	9.3	2	2.6
Shaanxi	38.06	24.9	17.5	5.3	9.6
Gansu	49.7	38.4	22.7	5.7	8.9
Qinghai	35.52	22.4	17.7	0.5	1.2
Ningxia	28.39	24.7	18.5	0.7	1.4
Xinjiang	45.64	22.3	27.4	1.4	4.8
China	22.25	13.9	6.3	100	100

Table 2.6 Urban poverty measures

	US$1.00 per day			US$1.50 per day			Official poverty line			World Bank
Year	P_0	P_1	P_2	P_0	P_1	P_2	P_0	P_1	P_2	P_0
					(percent)					
1992	2.09	0.45	0.17	13.74	2.6	0.82	2.48	0.51	0.18	0.83
1994	2.73	0.47	0.16	13.18	2.77	0.91	2.9	0.53	0.17	0.86
1995	1.65	0.36	0.12	10.27	1.98	0.64	1.68	0.29	0.08	0.61
1996	1.69	0.27	0.07	8.41	1.67	0.53	1.76	0.27	0.07	0.46
1997	2	0.42	0.14	9.21	2.06	0.71	2.44	0.45	0.14	0.53
1998	2.06	0.3	0.08	8.86	1.88	0.6	2.13	0.32	0.08	0.98
Annual growth rate (percent)										
1992–95	−7.58	−7.17	−10.96	−9.25	−8.68	−7.93	−12.17	−17.15	−23.69	−9.76
1996–98	10.41	5.41	6.9	2.64	6.1	6.4	10.01	8.87	6.9	45.96
1992–98	−0.24	−6.53	−11.81	−7.05	−5.26	−5.07	−2.5	−7.47	−12.8	2.81

Source: Except for the last column, all figures were calculated by Fang, Zhang, and Fan (2002). No official national urban poverty line is available in China, with different cities publishing different figures. The city-specific official poverty lines are taken from *China Development Report* (1998). The poverty line of the World Bank measure is US$32.74 per month. The figures are available from http://www.worldbank.org/research/povmonitor/countrydetails/China.htm.

Notes: P_0 refers to the headcount index, which measures the spread of poverty; P_1 is the poverty-gap ratio, which measures the depth of poverty; and P_2 is the square poverty-gap ratio, which measures the severity of poverty.

government introduced some safety net programs in urban areas, and promoted the development of small and medium enterprises for job creation.

Table 2.6 reports poverty estimates for urban China.[12] Three poverty measures are used: (1) the headcount index (P_0) which measures the spread of urban poverty; (2) the poverty-gap ratio (P_1), which measures the depth of poverty and is sensitive to changes in the average income of the poor; and (3) the square poverty gap (P_2), which measures the severity of poverty and is sensitive to changes in the inequality of income distribution of the poor.[13]

The incidence of urban poverty declined from 1992 to 1995 and increased thereafter following the implementation of major urban reforms (Table 2.6).[14] The level of poverty is sensitive to the choice of the poverty line. Based on the US$1.00 per day poverty line, the proportion of urban poor varied between 1.65 percent and 2.73 percent over the 1992–98 period. Given that real expenditure per capita grew on average by 6 percent per annum between 1992 and 1998, the rate of

[12] For more details about urban poverty measures, refer to the section on data.

[13] These poverty measures are members of the Foster–Greer–Throbecke class of poverty measures, known as P_a, $a = 0, 1, 2$.

[14] There have been several efforts to measure the urban poverty rate in China. Very often they showed conflicting trends owing to the use of different datasets and poverty lines. For example, estimates of urban poverty rate based on SSB's official data indicate a decline from 5.8 percent to 3.4 percent between 1991 and 2000 (Asian Development Bank). Applying an adjusted consumer price index on the same official data, Khan and Riskin (2001) estimated three measures of urban poverty (broad, deep, and extreme) in 1988 and 1995. Their estimates show that urban poverty grew over this period with an increase of 19.4 percent for broad poverty, 86.4 percent for deep poverty, and 145.5 percent for extreme poverty.

reduction of the number of people living on less than US$1.00 a day is astonishingly low, suggesting that the rapid economic growth has not trickled down to the poor. On the other hand, estimates based on a US$1.50 per day poverty line show not only a higher proportion of urban poor in China, but also a sharp decline in the poverty rate over the period of study. The incidence of urban poverty declined from 14 percent in 1992 to about 9 percent in 1998. The high sensitivity of the poverty estimates to small changes in the poverty line implies that a large share of the urban non-poor is concentrated just above the US$1.00 per day poverty line. Thus, these households are extremely vulnerable to external shocks and to a fall into poverty.

No matter which poverty line is used, the depth (P_1) and severity (P_2) of urban poverty declined sharply between 1992 and 1996, increased between 1996 and 1997, and declined afterwards. For the period as a whole, the depth and severity of urban poverty declined substantially, indicating that the average income as well as the distribution of income has improved.

There are marked regional variations in the incidence of urban poverty in China. Table 2.7 presents the proportion of urban poor for China's coastal, central, and western regions. Three features are particularly noteworthy. First, the fraction of the urban population living under the poverty line was higher in the western region than in central and eastern China over the period of study. Using the US$1.50 per day poverty line, about 20 percent of the urban populations of western cities were poor during the sample period compared to less than 8 percent in the coastal cities. This finding is not surprising given that the western region is home to many of the country's worst performing state-owned heavy industries. Factory clo-

Table 2.7 Urban poverty by region: Headcount index

	US$1.00 per day	US$1.50 per day	Official poverty line
		(percent)	
Coastal region			
1992	2.20	7.15	1.61
1994	4.95	7.89	1.50
1995	1.75	5.54	0.77
1996	4.14	4.42	0.65
1997	2.22	3.04	0.63
1998	1.07	3.48	0.70
Central region			
1992	2.22	19.36	1.89
1994	3.49	16.04	3.00
1995	1.53	13.89	1.48
1996	1.73	8.53	2.13
1997	2.10	12.27	3.02
1998	2.49	11.69	2.79
Western region			
1992	6.71	20.46	5.91
1994	7.38	22.23	6.50
1995	5.86	16.39	4.52
1996	5.07	18.98	4.10
1997	6.63	20.21	6.27
1998	6.32	17.70	4.65

Source: Calculated by Fang et al. (2002).

sures and layoffs resulting from the reforms have exacerbated the incidence of urban poverty in the region. Second, the incidence of poverty is highly sensitive to the choice of poverty line, particularly in the western and central regions. For example, the US$1.50 per day poverty line shows a poverty rate of about 12 percent in the central region in 1998 compared with only 2.5 percent when the US$1.00 per day poverty line is used. This indicates that a large number of households is clustered just above the US$1.00 per day poverty line. Third, the drop in urban poverty was more pronounced in the coastal region than in the central and western regions. While the incidence of poverty fell by 51 percent in eastern China between 1992 and 1998, it declined by 39 percent in central China, and by only 13 percent in the western region. Moreover, on the basis of both the US$1.00 per day and the official poverty lines, the incidence of urban poverty increased in the central region and stagnated in the western region between 1992 and 1998. These results suggest that the government should adopt more proactive measures to target the urban poor in the central and western regions if it wants to eliminate poverty more effectively.

CHAPTER 3

Development of Road Infrastructure

This chapter reviews the development of road infrastructure in China and draws on data from various issues of *China Transportation Yearbook,* 1984–2003.

Government Policies

The development of China's transportation infrastructure has been shaped by the various policy and institutional reforms that took place in the country during the past 50 years. Under the centrally planned system that existed prior to the economic reforms, the Chinese government exercised its authority in all sectors of the economy, including all decisions pertaining to infrastructure development. The inward-looking development strategy, oriented toward heavy industrialization and self-sufficiency, promoted the expansion of the transportation network in northern China where heavy industries were located. These policies also promoted railways, which were viewed as a cost-effective means of transporting material from resources-rich provinces to industrialized provinces (Démurger 2001). However, despite these improvements, the development of the transportation infrastructure remained relatively unimportant in China's national development strategy during the pre-reform period (World Bank 1999). Consequently, China was poorly endowed with transportation infrastructure when the reforms began in 1978. There were only 97 kilometers of roads per thousand square kilometers of land in China in 1980, compared with 230 kilometers of roads per thousand square kilometers of land in India.[15] Benziger (1993) argued that official Chinese statistics may have understated the length of rural roads in China, as most village-to-market roads are not included in China's road statistics. The author re-estimated road density in China based on information from Hebei Province and found that China's road density in the 1980s was only 20–30 percent below India's.

Despite the priorities given to develop China's transportation infrastructure in the early years of the reform period, investments were low compared to those in the industrial sector (Démurger 1999). The fiscal policies aimed at controlling inflation appeared to be a limiting factor in infrastructure development. During the 1980s, the government curbed overheated demands for consumption and infrastructure investments by controlling rigidly the money supply and by implementing a tight credit policy (Démurger 1999). Thus China's infrastructure development did not keep pace with a soaring demand in the early 1980s. The growth in inter-

[15]Road density estimates for India are based on road length data reported by the Indiastat database and refers to the length of paved roads. For China, road length data are reported in the *Chinese Statistical Yearbook* (2001) and refer to the length of highways. For both countries, land area data are taken from the *World Bank 2003 Development Indicator* database.

regional trade following the reforms combined with the relatively low level of investments in infrastructure created transportation shortages and urban congestion (Démurger 2001). Since 1985, the government has geared up its investment in roads, particularly high-quality roads such as highways connecting major industrial centers in the coastal areas.

In the 1990s, investments in infrastructure became a national priority. The Chinese government invested massively in road construction to speed up the expansion of the road network in counties and towns, to improve the quality of roads, and to increase the mileage of expressways. Although more resources were initially devoted to the coastal regions, the Chinese government has more recently shifted its focus to the western region. Road projects are now an important part of China's strategy to develop the western region. In 2003, for example, more than 200 projects were launched to improve transportation in western China. The aim of these projects was to connect all counties with highways and expand the length of the road network.

Development of Roads

Table 3.1 shows the development of the length and density of the road network in China from 1952 to 2002. The mountainous

Table 3.1 Characteristics of roads in China

	Length of highway	Road density	Passengers	Freight traffic
	(1,000 km)	(km/1000 km^2)	(million persons)	(million tons)
1952	127	13.6	45.6	131.6
1957	255	27.3	237.7	375.1
1962	464	49.7	307.4	327.9
1965	515	55.2	436.9	489.9
1970	637	68.3	618.1	567.8
1975	784	84.0	1,013.5	725.0
1978	890	95.4	1,492.3	851.8
1979	876	93.9	1,786.2	3,710.4
1980	883	94.7	2,228.0	3,820.5
1981	898	96.2	2,615.6	3,636.6
1982	907	97.2	3,006.1	3,792.1
1983	915	98.1	3,369.7	4,014.1
1984	927	99.4	3,903.4	5,333.8
1985	942	101.0	4,764.9	5,380.6
1986	963	103.2	5,442.6	6,201.1
1987	982	105.3	5,936.8	7,114.2
1988	1,000	107.2	6,504.7	7,323.2
1989	1,014	108.7	6,445.1	7,337.8
1990	1,028	110.2	6,480.9	7,240.4
1991	1,041	111.6	6,826.8	7,339.1
1992	1,057	113.3	7,317.7	7,809.4
1993	1,084	116.2	8,607.2	8,402.6
1994	1,118	119.8	9,539.4	8,949.1
1995	1,157	124.0	10,408.1	9,403.9
1996	1,186	127.1	11,221.1	9,838.6
1997	1,226	131.5	12,045.8	9,765.4
1998	1,279	137.1	12,573.3	9,760.0
1999	1,352	144.9	12,690.0	9,904.4
2000	1,403	150.4	13,473.9	10,388.1
2001	1,698	182.0	14,028.0	10,563.1
2002	1,765	189.2	14,752.6	11,163.2

Source: China's National Bureau of Statistics, *Statistical Yearbook of China*, 2001, 2003.

Table 3.2 Length of roads by class

	Expressway	Class 1	Class 2	Class 3	Class 4	Substandard	Total
				(km)			
1980		196	12,587	108,291	400,060	367,116	888,250
1981		214	14,126	111,770	409,605	361,893	897,607
1982		231	15,665	115,249	419,149	356,669	906,963
1983		255	17,167	119,203	426,190	352,264	915,079
1984		328	18,693	124,011	437,329	346,365	926,726
1985		422	21,194	128,541	456,286	335,952	942,395
1986		748	23,762	136,790	476,410	325,059	962,769
1987		1,341	27,999	147,838	491,212	313,853	982,243
1988	147	1,673	32,949	159,376	503,126	302,282	999,553
1989	271	2,101	38,101	164,345	511,105	283,866	999,789
1990	522	2,617	43,376	169,756	524,833	287,244	1,028,348
1991	574	2,897	47,729	178,024	535,444	276,468	1,041,136
1992	652	3,575	54,776	184,990	542,942	269,772	1,056,707
1993	1,145	4,633	63,316	193,567	559,472	261,343	1,083,476
1994	1,603	6,334	72,389	200,738	580,336	256,421	1,117,821
1995	2,141	9,580	84,910	207,282	606,841	246,255	1,157,009
1996	3,422	11,779	96,990	216,619	619,258	237,721	1,185,789
1997	4,771	14,637	111,564	230,787	635,737	228,909	1,226,405
1998	8,733	15,277	125,245	257,947	662,041	209,231	1,278,474
1999	11,605	17,716	139,957	269,078	718,380	194,955	1,351,691
2000	16,285	25,219	177,787	305,435	791,202	363,919	1,679,847
2001	19,437	25,214	182,102	308,626	800,665	361,968	1,698,012
2002	25,130	27,468	197,143	315,274	817,911	382,296	1,765,222
Growth rate (percent)							
1980–2002	44.4	28.5	14.0	5.1	3.2	–1.3	2.9

Source: China's National Bureau of Statistics, *Statistical Yearbook of China* (various issues).
Note: Growth rate for expressways is for 1988 to 2002.

topography in many parts of China has historically hindered the development of roads. In 1952, China's transportation infrastructure included only 126.7 thousand kilometers of roads, corresponding to a road density of 13.6 kilometers per thousand square kilometers of land. In that year, the road network handled about 132 million tons of goods and 45 million passengers, representing 41 and 19 percent of the total volume of goods and passengers conveyed by Chinese transportation system, respectively. By 2002, the length of roads had increased to 1.77 million kilometers, carrying some 14.7 billion passengers and 11.1 billion tons of goods. This implied a road density of 189 kilometers per square kilometer of land, a 14-fold increase over 1952. In the mid-1980s, the government emphasized the construction of high-quality roads. The construction of expressways has expanded substantially in China since the country completed its first expressway project in 1988. The length of expressways increased from 147 kilometers in 1988 to 25,130 kilometers in 2002, representing a more than 100-fold increase over 14 years (Table 3.2). Likewise, the length of class 1 highways increased rapidly, from 196 kilometers in 1980 to 27,468 kilometers in 2002, or an annual growth rate of 28.5 percent. Class 2 highways also expanded rapidly with an average annual growth rate of over 14 percent between 1980 and 2002. In contrast, the length of lower-quality roads expanded at much slower rate. The length of class 3 and class 4 roads increased by only 5.4 percent and 3.5 percent per annum from 1980 to 2002, whereas the length of substandard roads declined by 1.3 percent over the same period.

The expansion of the highway network contributed greatly to China's economic development. The State Planning and Development Commission estimated that highway construction has increased economic growth by up to 0.4 percentage points and has absorbed 4 to 5 million workers in the construction industry per year.

There are significant changes in the shares of different classes of roads in China over time. In 1980, high-quality roads (expressways and class 1 and class 2 roads) accounted for only 1.4 percent of the total road length, while low-quality roads (class 3, class 4, and substandard roads) accounted for 98.6 percent. In particular, class 4 and substandard roads constituted more than 85 percent of the total road length. By 2002, these shares had changed substantially, with high-quality roads accounting for 14.1 percent of the total length of roads while class 4 and substandard roads accounted for 68 percent of the total.

Large regional variations exist in the density and quality of road infrastructure in China (Table 3.3). The western region is poorly served by roads compared to the central and coastal regions. In 2002, there were only 166 and 66 kilometers of roads for every thousand square kilometers of land in southwest and northwest China, respectively, compared to more than 460 kilometers per thousand square kilometers of land in the eastern and central regions. Among all provinces, Tibet and Qinghai are particularly poorly endowed with road infrastructure, with a road density of only 33 kilometers per thousand square kilometers of land. Road quality is also the worst in the western region. In southwest China, for example, high-quality roads (expressways and class 1 and class 2 roads) account for less than 6 percent of the road network compared to 20 percent in the northern and the eastern regions.

Road development has also been uneven between rural and urban areas. In rural China, currently about 184 towns and 54,000 villages, most of which are located in the western region, have no access to roads. Large-scale road projects have recently been launched to expand and improve the rural road network (Xinhua News Agency 2003).

Sources of Funding

Prior to the reform period, road projects were funded predominantly by domestic sources in China. These sources included government appropriations, profits from state-owned enterprises, and local government levies. The central government was accountable for the development of national roads, while the provincial and local governments were responsible for the provincial and local road networks (Démurger 2001). Under the centrally planned system, provincial and local governments typically received funds for infrastructure construction from the central government.

Following the economic reforms, the sources of funds for roads have increasingly diversified and now include not only funds from central and local governments, but also loans from international organizations and banks, as well as foreign capital. Another important change has been the issuance of long-term public bonds to finance infrastructure projects. Between 1998 and 2002, the government issued more than 660 billion yuan in bonds. These bonds were issued to state-owned banks, such as the Industrial and Commercial Bank of China and the Agricultural Bank of China, and were assigned to projects aimed at:

- Infrastructure investment in agriculture, forests, water conservancy, and the environment.
- Construction of highways, railways, airfields, ports, and telecommunication projects.
- Environmental protection.
- Upgrading rural and urban electric networks.

As local governments were granted more autonomy in the post-reform period, they

Table 3.3 Length and density of roads by region, 2002

	Length of highways	Road density	Shares in total length of highways: Expressway, class 1 and 2	Shares in total length of highways: Below class 4
	(km)	(km per 1,000 km²)	(percent)	
North	146,745	392.2	19.7	8.5
Beijing	14,359	854.3	18.2	2.9
Tianjin	9,696	857.7	22.1	5.9
Hebei	63,079	332.0	21.4	14.4
Shanxi	59,611	382.1	17.9	4.0
Northeast	152,192	193.3	17.7	5.3
Liaoning	48,051	329.3	27.9	0.6
Jilin	41,095	219.3	16.0	6.5
Heilongjiang	63,046	138.9	11.0	8.2
East	368,500	463.3	20.5	15.5
Shanghai	6,286	991.4	30.0	4.2
Jiangsu	60,141	586.2	25.6	16.9
Zhejiang	45,646	448.4	20.1	6.3
Anhui	67,547	483.9	12.8	9.1
Fujian	54,155	446.1	11.9	23.9
Jiangxi	60,696	363.7	12.7	40.6
Shandong	74,029	472.4	35.4	0.2
Central	372,061	478.4	16.6	24.3
Henan	71,741	429.6	23.8	7.3
Hubei	86,098	463.1	14.9	21.8
Hunan	84,808	400.4	7.5	54.7
Guangdong	108,538	609.8	21.3	10.0
Hainan	20,876	596.5	10.1	43.6
Southwest	391,790	166.2	5.6	38.1
Chongqing	31,060	376.9	12.7	29.4
Sichuan	111,898	230.7	10.4	33.9
Guizhou	44,220	251.1	5.8	31.0
Yunnan	164,852	418.4	1.9	34.7
Tibet	39,760	32.6	1.5	79.0
Guangxi	56,297	237.8	10.7	25.1
Northwest	277,637	65.7	10.5	18.2
Inner Mongolia	72,673	61.6	9.2	13.3
Shaanxi	46,564	226.5	12.6	12.4
Gansu	40,223	88.4	11.5	23.4
Qinghai	24,003	33.3	12.8	23.0
Ningxia	11,245	170.4	21.4	1.5
Xinjiang	82,929	51.8	7.8	24.2
National total	1,765,222	184.7	14.1	21.7

Source: China's National Bureau of Statistics, *Statistical Yearbook of China,* 2003.

Table 3.4 Highway funding by sources

Year	Central government			Local government				Foreign capital
	Total	(1)	(2)	Total	(3)	(4)	(5)	
					(percent)			
1998	7.1	6.6	0.5	87.8	36.0	5.3	46.5	5.1
1999	6.7	5.9	0.8	89.7	36.3	4.9	48.6	3.6
2000	7.3	6.6	0.7	88.8	36.0	4.4	48.5	3.9
2001	12.0	8.9	3.1	85.1	40.7	3.6	40.9	2.9
Average	8.3	7.0	1.3	87.9	37.3	4.6	46.1	3.8

Source: China highway transport statistics collection (1998–2000).
Note: Numbers in the table represent the following variables: (1) ministry special funds; (2) central fiscal special funds; (3) domestic loans; (4) local fiscal funds; and (5) self-raised funds and others.

became responsible for most of the infrastructure projects financed by bonds. Local governments applying for expressway construction projects are now required to raise 35 percent of the cost themselves from their own revenue (including tolls) and by selling bonds. The remaining 65 percent of the cost is funded through bank loans. In the past, banks were reluctant to provide loans for road projects. Investment in roads, especially in highways and expressways, has proved to be beneficial in recent years, however. Consequently, banks are now more eager to fund road projects, especially in the eastern region. In western China, the situation is quite different. Highway and expressway investments are less profitable. As fewer cars use highways and expressways, local governments cannot get enough revenue from tolls to pay for road maintenance and to repay principal and interest.

Table 3.4 shows the breakdown in highway investments by source of funds. Between 1998 and 2001, local governments contributed the most to highway investments (88 percent), followed by the central government (8.3 percent) and foreign capital (3.8 percent). Two categories of funds make up the lion's share of highway investment: domestic loans and self-raised funds by local governments, which together accounted for more than 80 percent of the total investment in highways between 1998 and 2001.

The greater autonomy given to local governments also contributed to widening regional inequality as the capacity to raise funds to finance infrastructure projects depended on local government revenue (which in turn depended on the level of local economic activity) and the ability of local governments to negotiate higher contributions from the central government (Démurger 1999). The growing disparity in road provision across regions led the central government to launch major road construction projects in the central and western regions. As a result, the share of highway investment in eastern China declined from 54.8 percent in 1998 to 45.2 percent in 2001, whereas the corresponding shares in central and western China increased from 45.1 percent to 54.9 percent (Table 3.5).

Table 3.5 Regional shares of highway investments

Year	East	Central	West
		(percent)	
1998	54.8	23.9	21.2
1999	52.1	25.2	22.6
2000	49.2	26.8	24.0
2001	45.2	30.6	24.3
Average	50.0	26.8	23.1

Source: China highway transport statistics collection (1998–2000).

CHAPTER 4

Infrastructure Development and Poverty Reduction: A Literature Survey

A large literature exists on the impact of road investments on economic growth and poverty reduction. We classify this literature into three broad groups: (1) studies that examine the relationship between road investments and economic growth; (2) studies that look at the relationship between road investments and poverty alleviation; and (3) studies that focus on China. Whenever possible, we pay attention to studies that took into account the quality of roads in the analysis.

Economic Growth and Road Investments

To assess formally the contribution of road investments to production growth, a number of studies specify an aggregate production function that includes transportation infrastructure among the set of explanatory variables. Antle (1983), for example, estimated a Cobb Douglas production function for 47 developing countries and 19 developed countries. Infrastructure was specified as the gross national output from the transportation and communication industries per square kilometer of land area. Antle found a strong and positive relationship between the level of infrastructure and aggregate productivity. Ratner (1983), Binswanger et al. (1987), Aschauer (1989), Binswanger, Khandker, and Rosenzweig (1989), Baffes and Shah (1993), and Easterly and Rebelo (1993) also found transportation infrastructure was an effective factor of production.

There are several limitations to these studies. First, they do not take into account reverse causality. Reverse causality occurs if income growth increases the demand for infrastructure.[16] Ignoring reverse causality can lead to overestimation of the coefficients of the infrastructure variable in the production function. A second problem with these studies is the failure to take road quality into account. Road quality can vary greatly within a country, and different quality roads can act in different ways. Failure to discriminate among types of roads can also lead to biased estimates. The studies reviewed in the paragraphs that follow take these issues into consideration either jointly or in isolation.

Causality and Rates of Return to Roads

The only study we found that explored the direction of the causal links between infrastructure and productivity is that of Fernald (1999). Using data from 29 U.S. manufacturing industries

[16]For more information on reverse causality see Kessides (1993), World Bank (1994), and Canning and Bennathan (2000).

from 1953 to 1989, Fernald examined whether road investments lead to productivity growth or whether productivity growth entails greater road construction. His research findings suggest causation from roads to productivity, implying that the productivity decline in U.S. manufacturing after 1973 may have been a result of lower public spending on road infrastructure. Fernald's study also suggests that the marginal returns to road investments are not as high as commonly thought, primarily because road construction offers only a one-time increase in the level of productivity rather than a continuous series of impacts.

Using cointegration methods to circumvent reverse causality, Canning and Bennathan (2000) estimated the rates of return to paved roads for a panel of 41 countries over the past four decades. Canning found that the highest rates of return to road infrastructure occurred in countries with infrastructure shortages. Canning also analyzed whether physical capital, human capital, labor, and other infrastructure variables are complements or substitutes to roads. He found that the length of paved roads is highly correlated with physical and human capital. However, he observed that the marginal return to roads declines rapidly if the length of roads is increased in isolation from other inputs. Canning concluded that infrastructure investments are not sufficient by themselves to yield large changes in output. This finding is in line with that of Gannon and Zhi (1997), who also concluded that transport access is complementary to other services such as health and education.

Studies by Fan et al. (2000, 2002, 2004) in rural India, China, and Thailand also estimate the effect of infrastructure investments on economic growth and poverty. By estimating a system of equations, these studies explicitly account for the simultaneous effects of infrastructure investment in factor and product markets. Results from these studies consistently show the importance of road investments in promoting production growth and poverty reduction. In rural India, public investment in rural roads was found to have had the largest positive impact on agricultural productivity growth (Fan, Hazell, and Thorat 1999). In China and Thailand, road investments were found to have contributed significantly to growth in non-farm and total economic growth as well as to agricultural growth (Fan et al. 2002, 2004).

Explaining Productivity Differentials

Economic performance and income levels often differ greatly between regions within a country. Lagging regions (such as the northwest of China and northeast of Brazil) are commonly associated with poor infrastructure, which isolates local populations from educational, social, and economic opportunities and contributes to the rise of poverty traps. Deichmann et al. (2000), Nagaraj, Varoudakis, and Veganzones (2000), and Stephan (2000) have investigated the determinants of regional economic disparities. Nagaraj et al. (2000) assessed whether differences in the availability of physical, social, and economic infrastructure explained growth performance differentials among 17 Indian states from 1970 to 1994. Using instrumental variable estimation techniques to account for reverse causality, they found that a 10 percent increase in the road network (defined as kilometers of road per square kilometer of land) would lead to a 3.4 percent increase in income per capita. They also found that power consumption and health conditions are positively correlated with the availability of road infrastructure.

Disparities in the productivity of manufacturing firms between the southern states of Mexico and the rest of the country were the focus of the study by Deichmann et al. (2000). Specifically, Deichmann et al. aimed to assess the importance of differences in the quality of infrastructure in explaining productivity differentials. To account for quality, the authors developed a market access indicator defined as the size of the potential markets that can be reached from a particular point given the density and quality of the transportation network within that region. The econometric results presented in the study show that a 10 percent increase in market access increases labor productivity by 6 percent.

Stephan also found that differences in the level and quality of transportation infrastructure are significant in explaining differentials in regional economic performance. He studied the effects of road infrastructure on productivity for 21 French regions and 11 western German federal states and concluded that regional road infrastructure has a significant impact on regional output.

Access to Trade and Price Effects

The preceding studies have established links between the availability of transportation infrastructure and differences in economic performance among regions. Differences in regional economic performances can also be partly explained by a country or region's ability to trade as a result of better infrastructure. Limao and Venables (1999) elaborated on how the presence or absence of infrastructure influences access to trade. They constructed an infrastructure index that combines road, rail, and telecommunications densities. Using econometric methods, Limao and Venables studied the determinants of transportation costs. They showed that infrastructure is a significant determinant of transportation costs, and that when a region is landlocked, transport costs can increase by 50 percent. Using these findings along with detailed data on trade and transportation costs in Sub-Saharan Africa, they calculated that most of Africa's poor trade performance is the result of weak infrastructure. This finding concurs with similar findings by Delgado et al. (1995).

The availability and quality of road infrastructure can also influence food prices. Using survey data collected from itinerant traders, Minten and Kyle (1999) analyzed the causes of food price variation in Kinshasa, the capital of the former Zaire. They paid particular attention to the impact of distance and road quality on food price behavior and on the food collection system. Differences in road quality were accounted for by differentiating between paved roads and dirt roads. They reached the following conclusions. First, variations in food prices are significant across products and across regions. Second, transportation costs explain most of the differences in food prices among producer regions. Third, road quality was an important factor in determining transportation costs: transportation costs were on average two times greater on dirt roads than on paved roads.

Impacts of Road Investments on Poverty

The studies reviewed in the preceding section highlight the importance of roads in promoting economic growth and development. However, few of them provided information on the distributional and poverty impacts of road investments. To gain further insights into how road investments affect inequality and poverty reduction we turn to evidence from more micro-level studies. Most of the studies reviewed rely on analysis of household survey data. We also draw on evidence from project evaluation and appraisal reports on road projects.

Non-Farm Employment and Income Diversification

Road investments can help the poor in a number of ways, and one of the most important is through their impact on the rural non-farm economy. For example, rural road investments can promote the development of small non-farm enterprises, which in turn can increase the demand for rural labor. Using a reduced-form estimation technique and a panel dataset covering 85 districts in India over the period 1961–81, Khandker (1989) found that government investment in roads had a positive effect on crop output, rural non-farm employment, and agricultural wages, all of which were beneficial to the poor. Malmberg, Ryan, and Pouliquen (1997), Escobal (2001), and Fan and Rao (2002) have also explored the impact of roads on non-farm employment and the consequences for the poor. Malmberg et al. (1997) found that infrastructure investments contribute to economic growth in both farm and non-farm sectors, generating economic opportunities for the rural population in general, including the poor. Likewise, Fan and Rao (2002) concluded that non-farm employment became increasingly important in helping the poor during the post–green revolution period in many Asian countries. One of the consequences of greater non-farm employment is income diversification. Escobal (2001) established the link between roads and income diversification. He analyzed the determinants of rural household decisions to undertake off-farm activities in rural Peru. Using a Tobit doubled-censored estimation, Escobal showed that access to roads, along with other public assets such as rural electrification and education, is a significant determinant of income diversification. He also found that access to roads and other public assets raises the profitability of both farm and non-farm activities, but especially the latter.

Notwithstanding the positive economic impact of road infrastructure development in promoting the rural non-farm economy, the increasing importance of the rural non-farm sector is seen by many as a significant source of income inequality. Benjamin and Brandt (1999), for instance, ascribed off-farm employment as a main source of income inequality among households in northeast China. Similar findings were reached by Tsui (1991), Hussain, Lanjouw, and Stern (1994), Rozelle (1994), and Khan and Riskin (2001). Thus, the beneficial effect that non-farm activities have on rural income growth and poverty reduction may be offset by increasing inequality of income distribution.

Determinants of Poverty

Dercon and Krishnan (1998) and Kwon (2001) investigated the role of roads as one of several factors contributing to changes in the incidence of poverty. Kwon sought to identify the factors that contributed to the decline of poverty in 25 Indonesian provinces between 1976 and 1996. Her study showed that provinces with adequate road services were more likely to receive better irrigation services and produce more crops. People in these provinces seemed to have more job opportunities in the non-farm sector, either because they had easier access to labor markets or had more jobs available to them in the region. Using ordinary least-squares (OLS) and an instrumental variable estimation technique, Kwon found that roads have a significant impact on poverty alleviation. Her results also show that the impact of roads was greater in provinces with good access to roads than in provinces with poor access to roads.

Dercon and Krishnan (1998) used household data collected in rural Ethiopia in 1989, 1994, and 1995 to examine changes in poverty levels and to assess the factors driving the changes. By decomposing changes in poverty by subgroups of the population, they found that households with greater human and physical capital and with better access to roads had lower poverty levels. Dercon and Krishnan also noted that these factors reduce fluctuations in poverty over different seasons.

The Distribution of Benefits from Rural Roads

The studies reviewed in the preceding section confirm the importance of roads in poverty alleviation. However, the size and nature of the poverty effects and the distributional consequences remain unclear. In a rare study on this issue, Jacoby (2000) analyzed the distributional effects of rural roads in Nepal using household survey data. Jacoby developed a method for estimating benefits from road projects to households by assuming that lower transportation costs from better roads will be reflected in wages and farmland values. Based on these assumptions, Jacoby calculated the benefit that accrued to each household from a hypothetical road project and examined the distribution of these benefits across income classes. Using econometric estimation techniques, he found that providing improved road access to markets would generate substantial total benefits, a large share of which would be captured by poorer households. However, the benefits would not be large enough or targeted enough to significantly reduce income inequality.

Relying on a sample of 129 villages in Bangladesh, Ahmed and Hossain (1990) sought to estimate the impact of rehabilitated rural infrastructure. The authors established a strong positive effect of infrastructure on the incomes of the poor. Villages with better road access were significantly better off. For example, the use of fertilizer was 92 percent higher in these villages than in those with poorer infrastructure facilities. Ahmed and Hossain estimated that infrastructure endowment increased household income by 33 percent, almost doubled wages, and increased income from business and industries by 17 percent. In sum, better infrastructure was associated with greater agricultural output, higher incomes, better indicators of access to health services, and greater wage income opportunities. The authors concluded that the development of rural infrastructure has important implications for the alleviation of poverty.

To assess how road investments benefit the poor, Songco (2002) surveyed the impacts of rural infrastructure investments on household welfare. In addition to a literature review, Songco conducted a field survey in two provinces of the Central Highlands region of Vietnam to assess how poor households perceived benefits from upgrading low-grade roads to year-round access. The benefits identified by households and by local authorities are numerous and include improved mobility, reductions in the price of goods, and the elimination of health hazards from dusty roads. Moreover, Songco noted that for the poorest households the perceived impacts are mostly social (such as year-round access to school for children) rather than economic benefits. The rural poor acknowledged the importance of road improvements but indicated that interventions in other areas such as expanded credit opportunities are also important for improving household welfare. Songco cautioned that these results are specific to the Central Highlands of Vietnam and should not be extrapolated to other regions.

Another study that assessed the impact of a specific road project was undertaken by Khandker, Levy, and Filmer (1994). They reviewed the impact of a road project financed by the World Bank in Morocco and found increases had occurred in agricultural production and land productivity as well as in the use of agricultural inputs and extension services. The road project also led to a shift toward the production of high-value crops and an increase in off-farm employment opportunities. On the social front, benefits included improvements in access to health services and increased attendance at schools.

As discussed in Riverson, Gaviria, and Thriscutt (1991), Fishbein (2001), and Mahapa and Mashiri (2001), rural road projects do not always improve the well-being of local communities and help the poor. Mahapa and Mashiri assessed the impact of a road upgrading project in the village of Tshitwe in the Northern Province of South Africa. They surveyed about 140 households and found that the road improvement project was not cost efficient and failed to improve land productivity, off-farm employment, or to shorten the travel time to reach markets and other socioeconomic services. Mahapa and Mashiri also noted that road maintenance was neglected, as people did not receive the necessary training. Fishbein, in his review of the role of rural infrastructure in Africa's rural transformation process, found that the use of public funds has been inefficient and has left many people without basic access to roads. Riverson et al. reviewed 127 World Bank projects that involved rural roads in Sub-Saharan Africa. They found that the approaches used for planning and evaluation of rural roads had not paid sufficient attention to maintenance and had not fostered community participation. They also found that institutional problems were endemic to rural road projects. On the basis of these findings, Mahapa and Mashiri, Fishbein, and Riverson et al. stressed the importance of targeting interventions to local conditions as well as obtaining the participation of local communities for increasing the success of road projects. Howe (1981, 1997), Howe and Richards (1984), and Van de Walle (2000) reached similar conclusions.

Robinson (2001) pointed out that targeting and involving local communities in rural road projects—or decentralizing—is not always successful. Drawing on the literature and on field surveys in Nepal, Uganda, and Zambia, Robinson studied the effect of decentralizing transportation in developing countries. He found a number of constraints to successful devolution, including the lack of local government powers to exercise political influence, insufficient financial resources, and lack of management capability. He also found little evidence that existing decentralized systems address the needs of the rural poor. Robinson concluded that increased participation of the poor in the planning, financing, and implementation processes is important. Devres (1981), in a comprehensive survey of the literature on the socioeconomic and environmental impact of roads in developing countries, found that the larger and wealthier farmers are more likely to take advantage of new inputs, better technology, and extension services, and to respond to new market opportunities following road improvements.

Impact of Road Investments in China

The literature on the impact of road investments on economic growth and poverty reduction in China is comparatively sparse. Using provincial level data, Démurger (1999) and Felloni et al. (2001) assessed the consequences of infrastructure investments on production and productivity in China. Démurger emphasized the role of infrastructure endowments—that is, the length of railways, roads, and inland waterways per square kilometer of land—to explain growth performance differentials across provinces. Démurger's econometric results showed that cross-sectional differences in transportation infrastructure contribute significantly to the observed variation in growth performance among provinces. She also found a concave relationship between infrastructure endowment and economic growth. This suggests that expanding the transportation network will promote economic development in provinces with poor infrastructure endowment. On the other hand, upgrading or improving the quality of infrastructure may be more suitable for provinces with better transportation infrastructure. Hence, Démurger concluded that policies supporting infrastructure improvements could have substantial impacts in reducing disparities in the level of per capita income among Chinese provinces.

While Démurger assessed the overall economic impact of roads, Felloni et al. focused on the agricultural sector. Echoing Démurger's findings, Felloni et al. (2001) showed that the density of roads per hectare of agricultural land has a significant and positive effect on agricultural production and on land and labor productivity. Given that roads and energy are central to technology diffusion and production intensification and for facilitating access to the input and output markets, Felloni et al. argued that the availability of roads and electricity is crucial to the modernization of Chinese agriculture. These two studies, however, fail to consider the impact of road investments on poverty reduction as well as the simultaneous effects of infrastructure investments on factor and product markets.

Fan et al. (2002) take a more comprehensive approach to the problem. Using a model consisting of a system of equations to account for endogeneities, Fan et al. quantify the effects of rural infrastructure on growth and poverty reduction in rural China between 1970 and 1997. The authors found that public investments in roads, together with investments in education and agricultural research, helped to reduce rural poverty and regional inequality. Investments in roads also contributed to growth in agricultural production.

In contrast to the preceding studies, Lin and Song (2002) focused on the urban sector. Using data for 189 Chinese cities from 1991 to 1998, they found that an increase in paved roads is positively and significantly related to growth in GDP per capita in urban areas. Benziger (1996) provides interesting evidence on the linkages between the urban and rural sectors. Benziger tested whether greater access to infrastructure and to urban markets increases the intensity of input use and productivity in the rural sector in the province of Hebei. His econometric results show that road density and distance to the nearest city positively affect the use of fertilizer per unit of land, machinery per worker, and land and labor productivity.

Very little literature is available in Chinese on the impact of road investments in China. The impact of highway construction in China was the topic of investigations by Zhu (1990) and Liu (1999). Zhu observed that different phases of a highway project have different impacts. The construction period not only creates tremendous work opportunities, but also improves the skill of local people employed on the project. In the post-construction period, Zhu found that highways promote the development of goods production in poor regions, increase the volume of trade, reduce transportation costs, and improve social services. Liu reported the findings of a study on the macroeconomic impact of highway investment. Highway construction was found to have significant impacts on the economy by promoting employment and increasing farm incomes: for every 100 million yuan invested in highways, total output increased by 300 million yuan and created 7,000 jobs in the highway construction sector and related industries. On the other hand, Zhou (2001) showed that corruption in China acts to reduce the economic efficiency of public investments.

Summary

Despite differences in methodology, in research objectives, and in temporal and spatial coverage, the reviewed studies generally support the hypothesis of favorable impacts of roads on production and productivity, as well as on poverty alleviation. They also suggest that road investments can contribute to spatial inequities among regions. However, our survey also reveals several shortcomings in the literature, especially in its relevance to China. First, most studies focused on rural poverty. Until the early 1990s, poverty in China was considered largely a rural phenomenon and the rural poor were the focus of anti-poverty policies. Urban poverty became part of the political agenda only in the past decade as China shifted to a market economy and enjoyed rapid economic development. Second, nearly all the reviewed studies failed to take into account road quality in their specification. While the total length or density of road is a useful indicator of the road infrastructure available in a country, it is important to account for quality as different types of roads can have different economic returns and different impacts on poverty.

CHAPTER 5

Conceptual Framework and Model

Our conceptual framework is formulated to test the hypothesis that infrastructure investments in China promote economic development and growth and help reduce poverty. Poor residents in urban areas are hypothesized to benefit from increased employment opportunities and higher wages brought about by economic growth. Infrastructure investments in rural areas are hypothesized to affect poverty through various channels.[17] Infrastructure investment increases agricultural productivity, which in turn directly increases farm incomes and helps reduce rural poverty. Higher agricultural productivity also helps to lower rural poverty by increasing agricultural wages and improving non-farm employment opportunities. Moreover, improved agricultural productivity often leads to lower food prices, which helps the poor since they are typically net buyers of foods. In addition to their productivity impact, infrastructure investments directly increase rural wages, non-farm employment, and migration to urban or other rural regions. Understanding these different pathways can lead to useful policy insights for improving the effectiveness of government investments designed to promote growth and reduce poverty.

The Model

To systematically assess the impact of different types of public investment and different qualities of roads on both growth and poverty reduction, we developed a multi-equation model based on Fan et al. (2002). The first equation is an economy-wide labor productivity function[18]:

$$\text{GDPL} = f(\text{KSL, SCHY, ROAD1T, ROAD2T}), \tag{1}$$

where GDPL is the gross domestic product per worker measured in 1980 prices and KSL is capital stock per worker.[19] The variable SCHY is average years of schooling of the general population 15 years or older. To capture the impact of different types of roads, we use ROAD1T to represent the higher-quality roads per worker, that is, sum of length of expressway and class 1 and class 2 roads divided by the total number of workers, and ROAD2T to represent lower-quality roads, that is, sum of length of class 3, class 4, and substandard roads

[17] For more details see Fan et al. (2002).

[18] For definitions of variables used, refer to Table 6.1.

[19] The employment data are based on labor force surveys and represent employment by principal occupational category. Ideally, the actual labor inputs for various sectors should be used as a person can engage in multiple occupations, especially in rural areas. Unfortunately, these data are not readily available.

divided by total number of workers. To control for other factors not included in the equation, both year and provincial dummies are added.[20]

The second, third, and fourth equations represent labor productivity in urban, agricultural, and rural non-farm sectors. Agricultural labor productivity and non-farm labor productivity are functions of inputs as well as infrastructure and other public investment variables. Similarly, urban labor productivity growth is modeled as a function of urban inputs (labor and capital), infrastructure development, education, and other public investment variables in the urban sector. Different types of roads (by class) are included as separate variables in the production functions:

$$\text{UGDPL} = f(\text{UKSL, USCHY, ROAD1U, ROAD2U}) \quad (2)$$

$$\text{AGDPL} = f(\text{LANDP, FERTP, MACHP, RDS, IRRIP, RSCHY, RTRP, RELECP, ROAD1A, ROAD2A}) \quad (3)$$

$$\text{NFGDPL} = f(\text{NFKSL, RSCHY, ROAD1NF, ROAD2NF}), \quad (4)$$

where UGDPL, AGDPL, and NFGDPL are labor productivity in urban, agricultural, and rural non-farm sectors, respectively; UKSL and NFKSL are capital stocks in the urban and rural non-farm sectors, respectively; RSCHY and USCHY are average years of schooling for rural and urban residents, respectively. LANDP, FERTP, MACHP, RELECP, RTRP, and RDSP are land input, fertilizer use, machinery input, rural electricity consumption, number of rural telephone sets, and agricultural research (measured in stock terms) all expressed on a per agricultural laborer basis. IRRIP is the percentage of arable land under irrigation. ROAD1U and ROAD2U are the lengths of high- and low-quality roads per worker, respectively, in the urban sector; ROAD1A and ROAD2A are the lengths of high- and low-quality roads per worker, respectively, in the agricultural sector; and ROAD1NF and ROAD2NF are the length of high- and low-quality roads per worker, respectively, in the rural non-farm sector.

Agricultural prices are modeled as a function of agricultural GDP per worker (supply-side factor), and urban GDP per worker (demand-side factor):

$$\text{APRICE} = f(\text{AGDPL, UGDPL}), \quad (5)$$

where APRICE are real agricultural prices deflated by consumer price index. A world or border price variable is not included because China's food staple markets were closed during most of the time period of analysis.[21]

The next two equations model the determinants of rural and urban poverty.[22] The urban poverty equation is a function of labor

[20]Canning and Bennathan (2000) argued that a reverse causality may exist between GDP growth and infrastructure development. We followed Canning's approach by using one lead and two lags of differences of independent variables (KSL, ROAD1T, ROAD2T, and SCHY). Only the capital variable shows a strong reverse causality (i.e., the coefficients of one lead and two lags are significant). This may be due to the fact that region and year dummies may have wiped out the potential endogeneity effect.

[21]When a border price variable was included, it proved to be statistically insignificant.

[22]We also used a specification similar to that of Fan et al. (2002). That is, we estimated the poverty equation as a function of growth in agricultural labor productivity, rural wages, the percentage of rural non-farm employment, and the terms of trade of agricultural prices relative to non-agricultural prices. Agricultural labor productivity, rural wages, and non-farm employment are modeled as functions of public investment variables such as education, infrastructure, and agricultural R&D, together with other variables. The current specification allows us to calculate the economic returns and poverty reduction impact in both rural and urban sectors.

productivity in the urban sector, the Gini coefficient of income distribution of urban residents, and the agricultural terms of trade, controlling for other factors. Rural poverty is modeled as a function of growth in agricultural productivity, non-farm labor productivity, urbanization, and the agricultural terms of trade:

$$\text{UPOVERTY} = f(\text{UGDPL, UGINI, APRICE}) \quad (6)$$

$$\text{RPOVERTY} = f(\text{AGDPL, NFGDPL, APRICE, URBANP}), \quad (7)$$

where RPOVERTY and UPOVERTY are rural and urban poverty measured as the percentage of the relevant population under the poverty line, URBANP is the percentage of urban people in the total population, and UGINI is the urban Gini coefficient of per capita expenditure.[23]

Marginal Impact

The marginal impact of roads on growth in GDP, agricultural GDP, and urban GDP can be derived as:

$$\text{dGDP/dROADS} = \partial\text{GDPL}/\partial\text{ROADSL} \quad (8)$$

$$\text{dUGDP/dROADS} = \partial\text{UGDPL}/\partial\text{ROADSUL} \quad (9)$$

$$\text{dAGDPL/dROADS} = \partial\text{AGDPL}/\partial\text{ROADSAL} \quad (10)$$

$$\text{dNFGDP/dROADS} = \partial\text{NFGDPL}/\partial\text{ROADSNFL}. \quad (11)$$

Here ROADSL can be either high or low quality of roads. The coefficient of the length of roads per worker in the labor productivity function is the same as the coefficient of the length of roads in the GDP function when constant returns to scale are assumed. The marginal return per unit of length of roads is simply dGDP/dROADS* (GDP/ROADS).

Similarly, the marginal impact of roads on rural poverty can be derived as

$$\begin{aligned}\text{dRPOVERTY/dROADS} = {} & (\partial\text{RPOVERTY}/\partial\text{AGDP})(\partial\text{AGDP}/\partial\text{ROADS}) \\ & + (\partial\text{RPOVERTY}/\partial\text{NFGDP})(\partial\text{NFGDP}/\partial\text{ROADS}) \\ & + (\partial\text{RPOVERTY}/\partial\text{APRICE})(\partial\text{APRICE}/\partial\text{AGDP})(\partial\text{AGDP}/\partial\text{ROADS}) \quad (12)\end{aligned}$$

The first term on the right-hand side measures the impact of agricultural growth on rural poverty reduction while the second term captures the impact on rural poverty of improvements in rural non-farm GDP resulting from investment in roads.[24] The last term measures the impact on poverty resulting from changes in agricultural prices induced by increased agricultural production.

The marginal impact of improved roads on urban poverty is derived as

$$\begin{aligned}\text{dUPOVERTY/dROADS} = {} & (\partial\text{UPOVERTY}/\partial\text{UGDPP})(\partial\text{UGDPP}/\partial\text{ROADS}) + \\ & (\partial\text{UPOVERTY}/\partial\text{APRICE})(\partial\text{APRICE}/\partial\text{AGDPP})(\partial\text{AGDPP}/\partial\text{ROADS}). \quad (13)\end{aligned}$$

The first term of equation (13) is the impact of improved roads on urban poverty through urban growth. The second term captures the impact on urban poverty of lowered food or agricultural prices from increased production induced by road investments.

[23]Ideally, the rural Gini coefficient should also be included in the rural poverty equation, but the data on such a measure are available only for two years and for selected provinces as estimated by Khan and Riskin (2001). The inclusion of both agricultural and non-agricultural GDP per labor may help to alleviate the problem to some extent.

[24]The terms are separated by "+."

CHAPTER 6

Data, Model Estimation, and Results

Data

The sources of the data used in this study are, unless otherwise indicated, official publications by the Chinese statistical agency, the National Bureau of Statistics. Most labor, capital, public expenditures, education, and infrastructure variables are available annually at the provincial level from the 1950s to 2003. But poverty data at the provincial level are available only for selected years in the 1980s and 1990s. Therefore, certain statistical procedures and estimation techniques have to be used to fill data gaps and maximize the estimation efficiency.

Rural Poverty

There are several estimates of rural poverty in China. Official statistics indicate that the number of poor declined to about 30 million by 2000 (MOA, *China Agricultural Development Report* 2001). A second source is the estimates from the World Bank (World Bank 2000), which are similar to China's official statistics. A third set of estimates, based on a much higher poverty line (Ravallion and Chen 1997), shows a far greater proportion of the total population subject to poverty, with a poverty incidence of 60 percent in 1978 and 22 percent in 1995. However, the declining trend of rural poverty in this last set of estimates is steeper than that in the official Chinese statistics. Finally, Khan (1997), using samples of the national household survey, obtained 35.1 percent for 1988 and 28.6 percent for 1995.[25] Although these poverty rates are higher than the official rates, they confirm the declining trend shown in the official statistics.

Xian and Sheng (2001) provide the most recent estimates of rural poverty by province using a more rigorous approach. They used a poverty line of 860 yuan in terms of per capita consumption for 1998, which is actually higher than the US$1.00 per day poverty line criterion commonly used by the World Bank.[26]

We use provincial level poverty data from official sources. Few scholars have reported their estimates by province. Khan estimated provincial poverty indicators (both headcount ratio and poverty gap index) for 1988 and 1995 using the household survey data. To test the sensitivity of our estimated results, we first used both official statistics and Khan's estimates.

[25]The dataset included 10,258 rural households in 1998 and 7,998 in 1995.

[26]This is equivalent to US$1.15 per day measured in purchasing power parity.

We obtained similar results largely because the two sets of poverty figures share similar trends. Our final results are based on the official data simply because poverty data by province are available for more years, specifically from 1985 to 1989, and for 1991 and 1996.

Urban Poverty

The urban poverty and income variables were constructed by Fan, Fang, and Zhang (2001b) from China's urban household survey, which is conducted annually by the National Bureau of Statistics to monitor changes in urban household expenditures and consumption. A total of 40,000–50,000 households were surveyed annually between 1992 and 1998. We were able to access up to 10 percent of the total sample, taken from one representative city in each province.

To obtain appropriate poverty measures, we first had to convert our chosen poverty lines (US$1.00, 1.50, and 2.00 per capita per day, measured in 1985 purchasing power parity [PPP]) into local currency at nominal prices. To do this, we first converted the poverty line from 1985 PPP dollars into Chinese currency based on the 1985 PPP exchange rate. Then we used the Chinese consumer price index to calculate the national poverty lines at current prices. Finally, provincial level poverty lines were calculated by adjusting for differences in the cost of living by province.

To measure urban poverty, we used the percentage of the urban population falling below the chosen poverty line measured in 1985 PPP. There are good reasons to use a higher poverty line when measuring urban poverty. One prominent reason is the much higher cost of living for urban than for rural residents. Consequently, in this study we use poverty lines of US$1.50 and US$2.00 per capita per day. This leads to significant increases in the estimated number of urban poor in 1998, from 6.32 million when using the US$1.00 poverty line to 27.17 million and 60.04 million, respectively, when using the US$1.50 and US$2.00 poverty lines.

One important characteristic of the urban poor in China is the high share of total consumption expenditure spent on food. If the US$2.00 per capita per day poverty line is used, then the urban poor spent about 58 percent of their total expenditures on food in 1998 compared to 50 percent for the average urban population. Clearly, the urban poor would suffer the most from higher food prices.

Agricultural and Non-Agricultural GDP

Both nominal GDP and real GDP growth indices for various sectors are available from *The Gross Domestic Product of China* (SSB 1997a).[27] Data sources and construction of national GDP estimates were also published by the State Statistical Bureau in *Calculation and Methods of China's Annual GDP* (SSB 1997b). According to this publication, the SSB used the United Nations standard System of National Accounts definitions to estimate GDP for 29 provinces for three economic sectors (primary, secondary, and tertiary) in mainland China for the period 1952–95. Since 1995, the *China Statistical Yearbook* has published GDP data every year for each province for the same three sectors. Both nominal and real growth rates are available from SSB publications.

---

[27]There have been numerous debates about the accuracy of GDP measures in China. Rawski (2001) claimed that China's GDP growth rate has been overestimated by a large amount. For example, between 1997 and 1998, the official statistics reported a 7 percent growth, while Rawski claimed only a 2 percent growth. However, many Chinese scholars rebuffed his assertion. In recent years, the consensus has been that China's GDP may have been overestimated, but the magnitude of overestimation is only around 1–2 percent per annum. The objective of this study is not to resolve this debate. The regional and year dummies added in our regression may have largely reduced the potential bias on our estimated parameters.

The implicit GDP deflators by province for the three sectors are estimated by dividing nominal GDP by real GDP. These deflators are then used to deflate nominal GDP for rural industry and services to obtain their GDP in real terms.

Labor

Labor input data for the primary, secondary, and tertiary sectors at the provincial level after 1989 can be found in SSB's *Statistical Yearbook* (various issues), while provincial labor data prior to 1989 are available in SSB (1990). Labor is measured in stock terms as the number of persons at the end of each year. For rural industry and services, prior to 1984, labor input data at the township and village level, but not at the individual household level, are available in SSB's *China Rural Statistical Yearbook.* The omission of individual-household, non-farm employment data will not cause serious problems, as the share of this category in rural employment was minimal prior to 1984. Urban industry labor is estimated by subtracting rural industry labor from total industry labor, and urban service labor is similarly estimated as total service labor net of rural service labor. The labor input for the non-farm sector is calculated simply by subtracting agricultural labor from total rural labor.

Capital Stock

Capital stocks for the agricultural and non-agricultural sectors in rural areas are calculated from data on gross capital formation and annual fixed asset investment. The SSB (1997a) published data on gross capital formation by province for our three sectors after 1978. Gross capital formation is defined as the value of fixed assets and inventory acquired minus the value of fixed assets and inventory disposed. To construct a capital stock series from data on capital formation, we define the capital stock in time t as the stock in time $t - 1$ plus investment minus depreciation:

$$K_t = I_t + (1 - \delta)K_{t-1}, \qquad (14)$$

where K_t is the capital stock in year t, I_t is gross capital formation in year t, and δ is the depreciation rate. *China Statistical Yearbook* (SSB 1995) reports the depreciation rate of fixed assets of state-owned enterprises for industry, railways, communications, commerce, and grain for the period 1952–92. We use the rates for grain and commerce for agriculture and services, respectively. After 1992, the SSB ceased to report official depreciation rates. For the years after 1992 we used the 1992 depreciation rates.

To obtain initial values for the capital stock, we used a procedure similar to Kohli (1982). That is, we assume that prior to 1978 real investment grew at a steady rate (r), which is assumed to be the same as the rate of growth of real GDP from 1952 to 1977. Thus,

$$K_{1978} = \frac{I_{1978}}{(\delta + r)}. \qquad (15)$$

This approach ensures that the 1978 value of the capital stock is independent of the 1978–95 data used in our analysis. Moreover, given the relatively small capital stock in 1978 and the high levels of investment, the estimates for later years are not sensitive to the 1978 benchmark value of the capital stock.

To obtain the capital stock for the urban industrial sector, capital stock for rural industry is subtracted from the total industry capital stock (or secondary industry as classified by the SSB). Similarly, the capital stock for rural services is subtracted from the aggregate service sector (or tertiary sector as classified by the SSB) to obtain the capital stock for the urban service sector. Finally, the capital stock for rural enterprises is the sum of capital stocks for rural industry and services.

Prior to constructing capital stocks for each sector, annual data on capital formation and fixed asset investment were deflated by a capital investment deflator. The SSB began to publish provincial price indices for fixed asset investment in 1987. For years prior to 1987, we use the national price index of construction materials to proxy the capital investment deflator.

Roads

Based on the expected use, function, and the number of vehicles passed per day, highways are classified into five categories: expressways and class 1 to class 4 highways. Expressways can be classified into four-lane, six-lane, and eight-lane expressways, with associated increases in their vehicle carrying capacity. The designed carrying capacities for expressways are designated as follows: 25,000–55,000 minibuses or their equivalent per 24 hours for four-lane expressways; 45,000–80,000 minibus equivalents per 24 hours for six-lane expressways; and 60,000–100,000 minibus equivalents per 24 hours for eight-lane expressways. Other highways typically have two lanes and are classified by class with the following designed capacities: 15,000–30,000 minibus equivalents per 24 hours for class 1 highways; 3,000–7,500 minibus equivalents for class 2 highways; 1500–3000 minibus equivalents per 24 hours for class 3 highways; and fewer than 1,500 minibus equivalents per 24 hours for double lanes, or fewer than 200 minibus equivalents for single lanes for class 4 roads.

Substandard roads are usually rural roads connecting counties with towns and those connecting towns and villages. They are often not paved but are usually passable even when raining.

The designed length of life for roads also varies by type of road. For expressways and class 1 roads, the designed life is usually 20 years; for class 2 roads, 15 years; and for class 3 and class 4 roads, 10 years. Substandard roads normally have a lifespan of less than 10 years.

Agricultural Research and Development Expenditure

Public investment in agricultural research and development (R&D) is accounted for in the total national science and technology budget. Several government agencies invest in agricultural R&D. Science and technology commissions at different levels of government allocate funds to national, provincial, and prefectural institutes, primarily as core support. Institutes use these funds mainly to cover researchers' salaries, benefits, and administrative expenses. Project funds come primarily from other sources, including departments of agriculture, research foundations, and international donors. Recently, revenues generated from commercial activities (development income) became an important source of revenue for research institutes. The research expenditures reported in this study include only those expenses used to directly support agricultural research. The data reported here are from Fan and Pardey (1997) and various publications from the Government Science and Technology Commission and the State Statistical Bureau. Research expenditures and personnel numbers include those from research institutions at national, provincial, and prefectural levels, as well as agricultural universities (only the research part).

When calculating returns to R&D investment, expenditures on agricultural research as well as extension at the national and subnational levels are used as total R&D spending. This implicitly assumes that research conducted at the national level affects each province's production in proportion to the province's research expenditures, and the impact of extension conducted in each province is proportional to the province's extension expenditures.

Education

We use the percentage of population with different education levels to calculate the average years of schooling as our education variable, assuming 0 years for a person who is illiterate or semi-illiterate, 5 years for primary school education, 8 years for a junior high school education, 12 years for a high-school education, 13 years for a professional-school education, and 16 years for college and above education. The population census and the Ministry of Education report education levels by province for individuals older than age 7.

Rural Electricity

Total rural electricity consumption data for both production and residential uses by province from 1970 to 2002 are available in various issues of SSB's *China Rural Statistical Yearbook* and MOA's *China Agricultural Yearbook.* In more recent years, the *China Rural Energy Yearbook* (MOA 1995–2003) began publishing the use of electricity separately for residential and production purposes by province. We use this newly available information to backcast the different uses by province for earlier years.

Rural Telephones

The number of rural telephones is used as a proxy for the development of rural telecommunications. The number of rural telephones by province is published in various issues of *China Rural Statistical Yearbook* (SSB), the *China Statistical Yearbook* (SSB), and the *China Transportation Yearbook* (Ministry of Transportation).

Model Estimation

We used double-log functional forms for all equations in the system. More flexible functional forms such as the translog or quadratic impose fewer restrictions on the estimated parameters, but many interaction coefficients are not statistically significant because of multicollinearity problems. Model estimates also proved sensitive to slight alterations in the sample period or to deletion of non-significant variables. The use of a double-log functional form was deemed preferable, as it imposes some restrictions on the parameters and reduces the number of parameters to be estimated. This increases the number of degrees of freedom available and the reliability of the estimated coefficients.[28] Regional dummies were added to all equations to capture fixed effects arising from regional differences in agroclimatic and social economic factors. Year dummies were also added to control for any macroeconomic polices that may have had similar impacts on each region.

The literature review in the previous section indicates that if reverse causality is not considered then the effects of road investment may be overstated. Fan, Hazell, and Thorat (2000) used a difference approach, that is, taking first differences for all variables before estimation, to minimize any potential bias from reverse causality in their study on India. However, the first difference approach may eliminate all long-term relationships between public capital and economic growth. In their study on China, Fan et al. (2002) used the level approach to preserve the long-term relationship between investment and growth. They minimize potential bias by estimating an equation system in which they endogenize the public capital variables. Furthermore, they argued that public capital variables used in the pro-

[28]Fuss, McFadden, and Mundlak (1978) argue that functional forms chosen should satisfy the following criteria: (1) the functional forms should contain no more parameters than are necessary to agree with the maintained hypotheses; (2) the parameters should have intrinsic and intuitive economic interpretations, and a functional structure; (3) the trade-off between the computational requirements of a functional form and the roughness of empirical analysis should be weighted carefully in the choice of a model; (4) the chosen functional form should be well behaved, and should be consistent with such maintained hypotheses as positive marginal products or convexity, within the range of observed data; and finally (5) the functional form should be compatible with the maintained hypothesis outside the range of the observed data. The double-log function seems reasonable when judged against these criteria.

duction function are usually the results of past investments over many years while output or productivity is a function of the current capital stock. Therefore, the reverse causality should not exist unless investors or policymakers make their decisions based on the growth potential of each region. However, regional dummies included in the model should minimize the potential bias from this regional targeting.[29]

Equations (1) to (7) form a recursive system. Since there are non-zeros for some off-diagonal terms, a systems approach to estimation is still needed. As rural poverty data are available only for seven years at the provincial level (1985–89, 1991, and 1996) and urban poverty data are available at the provincial level only for 1992 to 1998, a two-step procedure was used in estimating the full equations system. The first step involved estimating all the equations except for the poverty equations using the provincial-level data from 1982 to 1999 with a full information likelihood estimation technique. Then the values of the independent variables in both the rural and urban poverty equations at the provincial level were predicted using the estimated parameters. In the second step, we estimated the rural and urban poverty equations using the predicted values of the independent variables at the provincial level based on the available poverty data. The advantage of this procedure is that it fully uses the information available for all the non-poverty equations, thereby increasing the reliability and efficiency of the estimates and avoiding endogeneity problems that can arise with the poverty equations.

Estimation Results

Table 6.1 summarizes variable definitions and Table 6.2 presents the estimated equations. In this section and thereafter we use low-quality roads and rural roads interchangeably, as low-quality roads are found mainly in rural areas. Similarly, we use high-quality roads and urban roads interchangeably.

The results for equation (1) (GDP per worker) show that the capital stock and human and infrastructure investments are all statistically significant in determining China's overall labor productivity. Both types of roads, high- and low-quality, are statistically significant with elasticities of 0.036 and 0.165, respectively. This means that for each 1 percent increase in high-quality roads, the GDP per worker will grow by 0.036 percent, while for every 1 percent increase in low-quality roads, the GDP per worker will grow by 0.165 percent.

The results for equation (2) show that urban capital plays a dominant role in urban labor productivity growth, with an elasticity of 0.547. Both types of roads contribute to urban labor productivity growth, although low-quality roads have the larger elasticity. This may be because low-quality roads help rural laborers to migrate to urban centers, and also provide markets for urban industrial products. The average years of schooling factor has a larger elasticity than the road variables.

The estimated agricultural labor productivity equation (equation [3]) shows that arable land per worker, fertilizer use, rural electricity consumption, rural education, agricultural research, and low-quality roads are all statistically significant. But high-quality roads do not show any statistically significant impact on agricultural productivity.

All the included variables are significant in the equation for rural non-farm labor

[29]Using the data from rural India, Zhang and Fan (2001) tested the two directions of causality between productivity growth and road capital. To avoid the reverse causality of road development to productivity growth, they used an instrumental variable approach, and found that the coefficient of roads changed very little when compared to the original model. One of the reasons was that road capital such as length of roads at the current level is a result of past government investments.

Table 6.1 Definition of variables

Variables	Definition
AGDPL	Agricultural GDP per worker
APRICE	Terms of trade, measured as agricultural prices divided by a relevant non-agricultural GNP deflator
FERTP	Chemical fertilizer use per worker
GDPL	GDP per worker
ILLITE	Rural illiteracy rate
IRRIP	Percentage of total cropped area that is irrigated
KSL	Capital stock per worker
LANDP	Arable land per agricultural worker
NFGDPL	Non-farm GDP per worker
NFKSL	Capital stock per worker in the rural non-farm sector
RDSP	Agricultural research stock per worker
RELECP	Rural electricity consumption per agricultural worker
ROAD1A	Length of high-quality roads per agricultural worker
ROAD1NF	Length of high-quality roads per non-farm worker
ROAD1T	Length of high-quality roads per worker
ROAD1U	Length of high-quality roads per urban worker
ROAD2A	Length of lower-quality roads per agricultural worker
ROAD2NF	Length of lower-quality roads per non-farm worker
ROAD2T	Length of lower-quality roads per worker
ROAD2U	Length of lower-quality roads per urban worker
RPOVERTY	Percentage of rural population below poverty line
RSCHY	Average years of schooling of rural population 15 years and older
RTRP	Number of rural telephone sets per agricultural worker
SCHY	Average years of schooling of general population 15 years and older
UGDPL	Urban GDP per worker
UGINI	Gini coefficient of per capita expenditure for urban residents
UKSL	Urban capital stock per worker
UPOVERTY	Percentage of urban population below poverty line
URBANP	Percentage of urban population in total population

productivity (equation [4]), but the elasticity of low-quality roads is much larger than that of high-quality roads. Rural education has a particularly large elasticity of 1.875.

The estimated terms-of-trade equation (equation [5]) confirms that increases in agricultural production exert a significant downward pressure on agricultural prices, worsening the terms of trade for agriculture.

The results for the urban poverty equation (equation [6]) show that increases in urban GDP significantly reduce urban poverty, while increases in urban inequality (as measured by the Gini coefficient) or agricultural prices significantly worsen urban poverty.

For rural poverty, the estimated results (equation [7]) show that improvements in labor productivity in the agricultural and rural non-farm sectors contribute significantly to rural poverty reduction. However, agricultural prices and the extent of urbanization are not significantly correlated with rural poverty. Previous work has shown the importance of increased agricultural prices on rural poverty reduction. The insignificant coefficient of agricultural price variable may come from the fact that the changes in agricultural prices are virtually the same among all provinces, and the provincial dummies included in the equation may have captured this effect. Another factor could be the high

Table 6.2 Estimates of the equations system

(1)	GDPL	=		.512 KSL	+	0.036 ROAD1T	+	0.165 ROAD2T	+	0.299 SCHY			$R^2 = 0.664$
				(12.81)*		(2.68)*		(3.30)*		(2.71)*			
(2)	UGDPL	=		0.547 UKSL	+	0.043 ROAD1U	+	0.172 ROAD2U	+	0.386 SCHY			$R^2 = 0.914$
				(15.1)*		(3.36)*		(3.65)*		(2.62)*			
(3)	AGDPL	=		0.023 LANDP	+	0.032 IRRIP	+	0.026 MACHP	+	0.179 FERTP	–	0.001ROAD1A	$R^2 = 0.833$
				(0.22)*		(0.34)		(0.50)		(2.76)*		(–0.39)	
			+	0.203 ROAD2A	+	0.014 RTRP	+	0.105 RELECP	–	0.223 ILLITE			
				(2.12)*		(0.84)		(2.09)*		(–2.13)*			
(4)	NFGDPL			0.526 NFKSL	+	0.052ROAD1NF	+	0.374 ROAD2NF	+	1.875 RSCHY			$R^2 = 0.921$
				(15.03)*		(2.11)*		(6.02)*		(6.54)*			
(5)	APRICE	=	–	0.087 ADGPL	+	0.095 UGDPL							$R^2 = 0.941$
				(–3.55)*		(1.59)							
(6)	UPOVERTY		–	0.941 UGDPL	+	1.233 GINI	+	0.519 APRICE					$R^2 = 0.578$
				(–5.67)*		(3.60)*		(5.63)*					
(7)	RPOVERTY	=	–	1.637 AGDPL	–	0.761 NFGDPL	–	0.126 URBANP	–	0.236 APRICE			$R^2 = 0.99$
				(–2.17)*		(–2.28)*		(–1.26)		(–0.18)			

Notes: Region and year dummies are not reported. Asterisks indicate that coefficients are statistically significant at the 10 percent level. The coefficients for the technology, education, and infrastructure variables are the sum of those for past government expenditures.

correlation of agricultural labor productivity and price variables.

Marginal Returns of Roads per Kilometer

Using the estimated equations (1) to (7) in Table 6.2 and the derived equations (8) to (13), we derived the marginal returns for different types of roads in terms of economic growth and rural poverty reduction. Because of the double-log functional form used, the marginal returns depend on the values of key right-hand-side variables at a specific period or point in time. To enhance the contemporary relevance of the results, the marginal returns were evaluated for 2001 rather than for the sample means or some other earlier year. The year 2001 falls beyond the time series used to estimate the model (ending in 1999), so there is an implicit assumption that the estimated model relationships are robust and still held in 2001. We calculated the marginal returns for different types of investments by region and for China as a whole. We divided China into seven regions according to geographic location, agricultural production structure, and the level of economic development at the provincial level as shown in Table 2.5.

Total GDP

Table 6.3 shows the marginal impacts on total GDP, urban GDP, agricultural GDP, and rural non-farm GDP of another kilometer of high- and low-quality roads. One more kilometer of high-quality roads (average of expressway and class 1 and class 2 roads) yields more than 1.7 million yuan worth of total GDP. Interestingly, the returns to road investments deviate little from the mean across regions. The southeast region has the largest return (2.2 million yuan), followed by the south (1.8 million yuan) and the southwest (1.7 million yuan), while the northwest region has the lowest return (1.1 million yuan). Other regions fall in between. For low-quality roads, every additional kilometer yields 1.1 million yuan of total GDP on average, about 66 percent of the corresponding return from high-quality roads. Further, in contrast to those to high-quality roads, returns to low-quality roads show much larger

Table 6.3 Economic returns to additional length of roads

	High-quality	Low-quality
	(yuan per kilometer)	
Returns in total GDP		
Average	1,730,748	1,158,072
Northeast	1,573,205	1,326,067
North	1,576,821	1,605,599
Northwest	1,109,934	416,690
Central	1,605,763	891,822
Southeast	2,245,363	3,651,586
Southwest	1,726,213	457,053
South	1,786,413	1,391,885
Returns in urban GDP		
Average	1,104,335	682,088
Northeast	1,102,441	857,776
North	1,033,951	971,835
Northwest	673,675	233,456
Central	840,450	430,870
Southeast	1,494,834	2,244,016
Southwest	947,638	231,607
South	1,223,394	879,885
Returns in agricultural GDP		
Average	NS	285,399
Northeast	NS	237,030
North	NS	329,720
Northwest	NS	129,162
Central	NS	339,018
Southeast	NS	558,810
Southwest	NS	197,963
South	NS	326,284
Returns in rural non-farm GDP		
Average	729,893	1,032,245
Northeast	595,471	875,066
North	671,564	1,244,221
Northwest	440,485	345,300
Central	797,611	1,081,150
Southeast	1,042,353	2,941,662
Southwest	487,823	330,103
South	701,359	1,141,634

Notes: Except returns in agricultural GDP to high-quality roads, all estimates are statistically significant at the 10 percent level. The returns are calculated for 2001 by using 2001 data.

regional differences. An additional kilometer of low-quality roads produces more than 3.6 million yuan worth of GDP in the southeast, compared to only 0.4 million yuan in the northwest.

Urban GDP

The marginal impact of another kilometer of high-quality roads is 1.1 million yuan in terms of urban GDP. As with total GDP, the return to high-quality road investments in urban GDP varies little among regions. It ranges from 0.67 million yuan in the northwest to 1.49 million yuan in the southeast. Low-quality roads have a lower return on average (0.68 million yuan of urban GDP) and this ranges from 0.23 million yuan in the northwest and southwest to 2.2 million yuan in the southeast. In general, road investments have lower returns in the less developed areas.

Agricultural GDP

As high-quality roads have only a small and insignificant impact on agricultural GDP (equation [3]), we do not calculate their marginal returns. For low-quality roads, every additional kilometer generates 0.29 million of agricultural GDP. The highest return occurs in the southeast (5.6 million yuan) while the lowest return occurs in the northwest (1.3 million yuan).

Rural Non-Farm GDP

Low-quality roads yield higher marginal returns to rural non-farm GDP than high-quality roads. On average, every additional kilometer of high-quality roads yields 0.73 million yuan of rural non-farm GDP, while low-quality roads yield more than 1 million yuan. Not surprisingly, returns to both types of roads are highest in the southeast, while the lowest returns occur in the southwest and northwest.

Urban Poverty

The estimated marginal impact of road investments on urban poverty reduction tells a different story (Table 6.4). The highest returns occur in poor regions such as the northeast, northwest, and southwest. Each additional kilometer of high-quality roads lifts about 10 to 15 urban poor out of poverty in the northwest, northeast, and southwest; in the southeast and the south, fewer than 2 urban poor would be affected. For China as a whole, each additional kilometer of high-quality roads lifts about 6 urban poor out of poverty. Turning to low-quality roads, each additional kilometer

Table 6.4 Returns in poverty reduction to additional length of roads

	High-quality	Low-quality
	(number per kilometer)	
Returns in urban poverty reduction		
Average	5.53	3.61
Northeast	13.23	10.87
North	3.18	3.16
Northwest	14.84	5.43
Central	5.16	2.79
Southeast	1.79	2.84
Southwest	9.52	2.46
South	0.72	0.55
Returns in rural poverty reduction, official data		
Average	8.97	21.59
Northeast	4.36	13.22
North	6.99	25.60
Northwest	27.91	37.69
Central	4.36	8.72
Southeast	1.07	6.23
Southwest	36.63	34.90
South	2.94	8.25
Returns in rural poverty reduction, Xian and Sheng data		
Average	34.96	109.61
Northeast	16.52	53.65
North	55.14	226.30
Northwest	57.57	98.29
Central	32.92	97.75
Southeast	14.06	92.02
Southwest	91.75	135.77
South	10.79	39.81

Notes: All estimates are statistically significant at the 10 percent level. The returns are calculated for 2001 by using 2001 data.

raises about 4 urban poor above the poverty line. Low-quality roads have the largest impact on urban poverty in the northeast while the lowest impact occurs in the south.

Rural Poverty

Two sources of rural poverty data are used to calculate the effects of road investments on rural poverty: the poverty rate by province reported by Chinese official statistics (SSB) and the poverty rate estimated by Xian and Sheng (2001). Using the official data, Table 6.4 shows that 9 rural poor would be lifted above the poverty line in China by each additional kilometer of high-quality roads. The largest impacts arise in the southwest and northwest regions with 37 and 28 rural poor lifted above the poverty line, respectively. In contrast, each additional kilometer of low-quality road lowers rural poverty by about 22 people in China. Again, the largest poverty reduction effect occurs in the southwest, followed by the northwest.

If Xian and Sheng's poverty data are used, the marginal impact of high-quality roads on rural poverty increases to 35 poor people per additional kilometer.[30] The largest poverty reduction impact from high-quality road expansion takes place in the southwest, followed by the northwest and the north. In the southwest and the northwest, the poverty effects are particularly large, with 91 and 57 rural poor lifted above the poverty line per kilometer of road. Interestingly, the north has the largest marginal impact per kilometer of low-quality road followed by the southwest. Our estimates indicate that for the northern region, 226 rural poor would be lifted out of poverty for every additional kilometer of low-quality road. The marginal impacts of low-quality road investments on rural poverty are similar in the central and southeast regions, whereas the impact on rural poverty is the lowest in the south and the northeast.

Marginal Returns of Roads per Unit of Investment

Using the unit costs of constructing different types of roads in different regions, we can express the marginal economic and poverty effects of additional road investments on a unit cost basis. The unit costs for different types of road construction are shown by region in Table 6.5. The unit cost of expressway construction varies little among regions, with the highest cost occurring in the southeast and the lowest cost in the north. The high cost in the southeast may reflect the high cost of land while the lower cost in northern China may arise from its flat topography and lower cost of land. On the other hand, large regional variations are observed in the unit construction costs of lower-quality roads. For class 4 and substandard roads, the lowest cost is found in the southwest while the highest cost occurs in the south. We calculated the average unit cost for expressways and class 1 and class 2 roads as the average cost for high-quality roads while we treat the rest as low-quality roads. At the national level, high-quality roads cost six to eight times more than low-quality roads.

Two assumptions were made in estimating the total annual costs of roads per kilometer. First, we assume that high-quality roads last for 16 years and low-quality roads for 10 years.[31] Second, we assume the maintenance cost is equivalent to the annualized-capital cost. For example, the unit cost of construction of 1 kilometer of high-quality

[30]This calculation is based on the assumption that the estimated relationship between investment and poverty reduction holds in 2001. It also assumes that this relationship, estimated based on a lower poverty line (US$0.66 per day), holds for the higher poverty line (US$1.15 per day).

[31]The expressways and class 1 roads have 20 years of lifespan while class 2 has 15 years. We use length of these different types of roads as weights in calculating the weighted average of lifespan for high-quality roads. For low-quality roads, the lifespan is 10 years.

Table 6.5 Unit cost of construction by type of roads

	Expressway	Class 1	Class 2	Class 3	Class 4	Substandard	High-quality	Low-quality
					(1,000 yuan)			
Average	26,110	9,100	2,850	1,430	500	380	5,600	670
Northeast	29,780	10,380	3,070	1,540	540	400	5,810	930
North	20,650	7,200	3,330	1,670	580	440	5,310	660
Northwest	22,480	7,840	3,450	1,730	600	450	4,900	730
Central	25,940	9,040	2,550	1,280	450	340	4,690	340
Southeast	34,950	12,180	2,520	1,260	440	330	6,990	500
Southwest	23,440	8,170	1,280	640	220	170	4,530	150
South	26,870	9,360	3,730	1,870	650	490	6,930	460

Source: Estimated from Zhao's data (Ministry of Transportation).

roads is 5.6 million yuan. Since the maintenance and service cost is the same as the annualized capital cost, the total cost for 1 kilometer of high-quality roads is 11.2 million yuan.

Table 6.3 indicates that every kilometer of high-quality roads constructed in 2001 would increase total GDP by 1.73 million yuan. This effect is assumed to begin in the fifth year after the investment was made. Once this effect takes place, it is assumed to last for the life of the road, in this case 16 years. Using a 5 percent discount rate, the total present-day value amounts to 16.2 million yuan. This can now be expressed on a unit cost basis, or as a marginal benefit–cost ratio. For every yuan invested in high-quality roads, 1.45 yuan (16.2/11.2) of total GDP would be added as shown in Table 6.6.[32] In the case of low-quality roads, we assume that once the investment is made, it will begin to have an impact after 3 years and it will last for 10 years. Similar calculations are made for the poverty reduction effects. Table 6.4 shows that each additional kilometer of high-quality roads lifts 5.53 urban poor above the poverty line, and Table 6.7 shows that when expressed on a unit cost basis, for every 10,000 yuan, the poverty reduction effect in urban area is 0.08 (or 5.53/11.2 × 16/100). As both assumptions are very conservative, our estimated returns to road investments are almost certainly lower bounds.

Total GDP

For the country as a whole, the marginal benefit–cost ratio for high-quality roads was 1.45 in 2001 (Table 6.6). The southwest region has the highest return, followed by the central region, which are two relatively poor regions in China. The lowest returns occurred in the northwest and the south. The returns to low-quality roads are much higher. The average return to low-quality roads was 6.37 yuan for each yuan invested in China in 2001. This is more than four times larger than the return to investment in high-quality roads. The southeast region has the highest return, followed by the south and southwest, whereas the lowest return occurs in the northwest.

[32]Alston et al. (2000) developed an approach to convert a benefit–cost ratio to an internal rate of return or vice versa. It is assumed that the benefit stream is a perpetual annual flow, B, per year while the cost is a one-time spending, C at time t. Thus the net present value of B is: $PV(B)_t = \Sigma_{j=0}^{\infty} B_{t+j}/(1+i)^j \approx \Sigma_{j=0}^{\infty} B/(1+i)^j = B/i$ and the net present value of cost is $PV(C)_t \approx C_t = PV(B)_t$ (at IRR) $\approx B/IRR$. Therefore, the benefit–cost ratio is $BC_t = PV(B)_t \div PV(C)_t \approx (B/i) \div (B/IRR) = IRR/i$, where i is the discount rate, IRR is the internal rate of return, and BC is the benefit–cost ratio. Hence we can also approximate the IRR as BC^*i. If $BC > 1$, then IRR is always greater than i.

Table 6.6 Returns in total GDP to road investment

	High-quality	Low-quality
	(yuan per yuan)	
Returns in total GDP		
Average	1.45	6.37
Northeast	1.27	5.25
North	1.39	8.89
Northwest	1.06	2.09
Central	1.60	9.57
Southeast	1.50	27.09
Southwest	1.78	10.86
South	1.21	11.02
Returns in urban GDP		
Average	0.92	3.75
Northeast	0.89	3.40
North	0.91	5.38
Northwest	0.64	1.17
Central	0.84	4.62
Southeast	1.00	16.65
Southwest	0.98	5.50
South	0.83	6.96
Returns in agricultural GDP		
Average	NS	1.57
Northeast	NS	0.94
North	NS	1.83
Northwest	NS	0.65
Central	NS	3.64
Southeast	NS	4.15
Southwest	NS	4.70
South	NS	2.58
Returns in rural non-farm GDP		
Average	0.61	5.68
Northeast	0.48	3.47
North	0.59	6.89
Northwest	0.42	1.73
Central	0.80	11.60
Southeast	0.70	21.83
Southwest	0.50	7.84
South	0.47	9.04

Notes: Except returns in agricultural GDP to high-quality roads, all estimates are statistically significant at the 10 percent level. The returns are calculated for 2001 by using 2001 data.

Table 6.7 Returns in poverty reduction to road investment

	High-quality	Low-quality
	(number per million yuan)	
Returns in urban poverty reduction		
Average	8	27
Northeast	18	59
North	5	24
Northwest	24	37
Central	9	41
Southeast	2	29
Southwest	17	79
South	1	6
Returns in rural poverty reduction, official data		
Average	13	161
Northeast	6	71
North	11	193
Northwest	46	257
Central	7	127
Southeast	1	63
Southwest	65	1,127
South	3	89
Returns in rural poverty reduction, Xian and Sheng data		
Average	50	820
Northeast	23	289
North	83	1,705
Northwest	94	671
Central	56	1,427
Southeast	16	928
Southwest	162	4,386
South	12	428

Notes: All estimates are statistically significant at the 10 percent level. The returns are calculated for 2001.

Urban GDP

One surprising finding is the low urban GDP return of high-quality roads. For every yuan invested, high-quality roads yield 0.92 yuan of urban GDP for China as a whole. The benefit–cost ratio is about 1. This implies that building more high-quality roads will not result in economically meaningful returns to urban GDP. But for lower-quality roads, the average return is 3.75 yuan per yuan invested, more than four times the effects of high-quality roads. The southeast region has by far the highest return.

Agricultural GDP

High-quality roads do not have a statistically significant impact on agricultural GDP. For low-quality roads, every yuan invested yields 1.57 yuan worth of agricultural GDP. Again, the patterns among regions are similar to those estimated for total GDP and urban GDP. The southeast, southwest, and central regions have higher returns than the national average, while the northwest, northeast, and north have lower returns than the national average.

Rural Non-Farm GDP

For every yuan invested in high-quality roads, the average return in rural non-farm GDP in China is 0.61 yuan. In contrast, low-quality roads yield 5.68 yuan worth of non-farm GDP for every yuan invested in roads. Moreover, the marginal returns from high-quality roads differ little across regions, whereas large regional variations exist for low-quality roads. The highest return occurs in the southeast, while the lowest arises in the northwest.

Urban Poverty

For every 1 million yuan invested in high-quality roads, 8 urban poor would be lifted above the poverty line (Table 6.7). Low-quality roads are more beneficial to the urban poor, raising 27 urban poor above the poverty line for each million yuan invested in 2001. This larger impact arises primarily because low-quality roads, being mostly rural, induce a larger increase in national food production and hence reduce food prices more. Roads have a greater impact on urban poverty in the less developed western regions (both the southwest and northwest) and the northeast region.

Rural Poverty

For high-quality roads, every million yuan invested raises 13 rural poor above the official poverty line (Table 6.7). Again, low-quality roads are much more beneficial, raising 161 rural people out of poverty for every million yuan invested. For both high-quality and low-quality roads, the poverty impacts are largest in the southwest and northwest regions when the official poverty line is used. However, when Xian and Sheng's poverty line is used, the number of rural poor helped is much larger: for each million yuan invested in high- and low-quality roads, 50 and 821 rural poor, respectively, are raised above the poverty line. These effects are about four times larger than those estimated using the official poverty line. For high-quality roads, the largest impact is found in the southwest, followed by the northwest and the north. For low-quality roads, the largest impact also occurs in the southwest followed by the north, and then the central and southeast regions.

CHAPTER 7

Conclusions

China has been very successful in achieving rapid economic growth and poverty reduction in recent decades. Driving this success was a series of policy and institutional reforms and massive public investments in roads and other key infrastructures. The primary objective of this report has been to analyze the contribution of road investments to China's successful transformation.

Using provincial-level data for 1982–99, an analytical framework was developed that extends earlier work by Fan et al. (2002) by differentiating among roads of different quality, and by disaggregating the measured effects of road investments by rural and urban areas. The results show that road development, together with agricultural R&D, irrigation, education, electricity, and telecommunications, made significant contributions to economic growth and poverty reduction. But variations in the marginal impact of roads on growth and poverty reduction were large, both between different types of roads and between regions.

The most significant finding of this study is that low-quality (mostly rural) roads have benefit–cost ratios for national GDP that are approximately four times larger than the benefit–cost ratios for high-quality roads. Even in terms of urban GDP, the benefit–cost ratios for low-quality roads are much greater than those for high-quality roads. As far as agricultural GDP is concerned, high-quality roads do not have a statistically significant impact while low-quality roads are not only significant but also generate 1.57 yuan of agricultural GDP for every yuan invested. Investment in low-quality roads also generates high returns in rural non-farm GDP. Every yuan invested in low-quality roads yields more than 5 yuan of rural non-farm GDP.

In terms of poverty reduction, low-quality roads raise far more rural and urban poor above the poverty line per yuan invested than do high-quality roads.

Another significant finding of the study is the trade-off between growth and poverty reduction when investing in different parts of China. Road investments yield their highest economic returns in the eastern and central regions of China while their contributions to poverty reduction are greatest in western China (especially the southwest region). This implies different regional priorities depending on whether economic growth or poverty reduction is the most important goal for the country.

The results of this study have important implications for future road project investments. In the past, China has invested heavily in building expressways and intercity highways. These investments have been a major force in China's economic transformation during the 1980s and 1990s. However, as more and more investments are being poured into these projects, the marginal returns are beginning to decline, although they are still positive and economically sound. At the same time, low-quality roads or rural roads have received less attention than high-quality roads, and as a result their marginal returns are much larger today than the returns to high-quality roads. Low-quality roads also raise more poor people out of poverty per yuan invested than do high-quality roads, making them a win–win strategy for growth and poverty alleviation. The government should now consider giving greater priority to low-quality and rural roads in its future investment strategy.

References

Ahmed, R., and M. Hossain. 1990. *Developmental impact of rural infrastructure in Bangladesh.* IFPRI Research Report No. 83. Washington, D.C.: International Food Policy Research Institute.

Alston, J. M., C. Chan-Kang, M. C. Marra, P. G. Pardey, and T. J. Wyatt. 2000. *A meta-analysis of rates of return to agricultural R&D: Ex pede herculem?* IFPRI Research Report No. 113. Washington, D.C.: International Food Policy Research Institute.

Antle, J. M. 1983. Infrastructure and aggregate agricultural productivity: International evidence. *Economic Development and Cultural Change* 31 (April): 609–619.

Aschauer, D. A. 1989. Is public expenditure productive? *Journal of Monetary Economics* 23: 177–200.

Baffes, J., and A. Shah. 1993. *Productivity of public spending, sectorial allocation choices and economic growth.* Paper presented at the 1993 Annual Meetings of American Economic Association, Anaheim, California.

Benjamin, D., and L. Brandt. 1999. Markets and inequality in rural China: Parallels with the past. *American Economic Review* 89 (2): 292–295.

Benziger, V. 1993. China's rural road system during the reform period. *China Economic Review* 4 (1): 1–17.

———. 1996. Urban access and rural productivity growth in post-Mao China. *Economic Development and Cultural Change* 44: 539–570.

Binswanger, H., C. Yang, A. Bowers, and Y. Mundlak. 1987. On the determinants of cross-country aggregate agricultural supply. *Journal of Econometrics* 36 (1): 111–113.

Binswanger, H., S. Khandker, and M. Rosenzweig. 1989. *How infrastructure and financial institutions affect agricultural output and investment in India.* World Bank Working Paper Series 163. Washington, D.C.: World Bank.

Canning, D., and E. Bennathan. 2000. *The social rate of return on infrastructure investments.* Policy Research Working Paper Series 2390. Washington, D.C.: World Bank.

Deichmann, U., M. Fay, J. Koo, and S. V. Lall. 2000. *Economic structure, productivity, and infrastructure quality in southern Mexico.* Policy Research Working Paper Series 2900. Washington, D.C.: World Bank.

Delgado, C. L., J. Hopkins, and V. A. Kelly with P. Hazell, A. A. McKenna, P. Gruhn, B. Hojjati, J. Sil, and C. Courbois. 1995. *Agricultural growth linkages in Sub-Saharan Africa.* IFPRI Research Report No. 107. Washington, D.C.: International Food Policy Research Institute.

Démurger, S. 1999. *Differences in infrastructure investments: An explanation for regional disparities in China?* Presented at CEANA session at ASSA meeting, Boston, January 7–9, 2000.

———. 2001. Infrastructure development and economic growth: An explanation for regional disparities in China? *Journal of Comparative Economics* 29: 95–117.

Dercon, S., and P. Krishnan. 1998. *Changes in poverty in rural Ethiopia 1989–1995: Measurement, robustness tests and decomposition.* WPS/98-7. Oxford: Center for the Study of African Economics.

Devres Inc. 1981. *Socio-economic and environmental impacts of low volume rural roads—A review of the literature.* AID Program Evaluation Discussion Paper No. 7. Washington, D.C.: Office of Evaluation, Bureau for Program and Policy Coordination, Agency for International Development.

Easterly, W., and S. Rebelo. 1993. Fiscal policy and economic growth: An empirical investigation. *Journal of Monetary Economics* 32 (3): 417–458.

Escobal, J. 2001. The determinants of nonfarm income diversification in rural Peru. *World Development* 29 (3): 497–508.

Fan, S., and P. G. Pardey. 1997. Research, productivity, and output growth in Chinese agriculture. *Journal of Development Economics* 53: 115–137.

Fan, S., and N. Rao. 2002. Public investment and poverty reduction: A synthesis of issues, methods and major findings. Mimeo. Washington, D.C.: International Food Policy Research Institute.

Fan, S., P. Hazell, and S. Thorat. 1999. *Government spending, agricultural growth and poverty: An analysis of interlinkages in rural India.* IFPRI Research Report No. 110. Washington, D.C.: International Food Policy Research Institute.

———. 2000. Government spending, agricultural growth, and poverty in rural India. *American Journal of Agricultural Economics* 82 (4): 1038–1051.

Fan, S., L. Zhang, X. Zhang, and X. Ma. 2001a. Regional priorities of public investment in rural China: A county-level analysis. Report prepared for the project Priorities of Public Investments in Chinese Agriculture. Washington, D.C.: International Food Policy Research Institute.

Fan, S., C. Fang, and X. Zhang. 2001b. How agricultural research affects urban poverty in developing countries: The case of China. EPTD Discussion Paper 80. Washington, D.C.: International Food Policy Research Institute.

Fan, S., L. Zhang, and X. Zhang. 2002. *Growth, inequality, and poverty in rural China: The role of public investments.* IFPRI Research Report No. 125. Washington, D.C.: International Food Policy Research Institute.

Fan, S., S. Jitsuchon, and N. Methakunnavut. 2004. *The importance of public investment for reducing rural poverty in middle-income countries: The case of Thailand.* DSGD Discussion Paper No. 7. Washington, D.C.: International Food Policy Research Institute.

Fang, C., X. Zhang, and S. Fan. 2002. Emergence of urban poverty and inequality in China: Evidence from household surveys. *China Economic Review* 13 (4): 430–443.

Felloni, F., T. Wahl, P. Wandschneider, and J. Gilbert. 2001. *Infrastructure and agricultural production: Cross-country evidence and implications for China.* TW-2001-103. Pullman: Washington State University.

Fernald, J. G. 1999. Roads to prosperity? Assessing the link between public capital and productivity. *The American Economic Review* 89 (3): 619–638.

Fishbein, R. 2001. *Rural infrastructure in Africa: Policy directions.* Africa Region Working Paper Series No. 18. Washington, D.C.: World Bank.

Fuss, M., D. McFadden, and Y. Mundlak. 1978. A survey of functional forms in the economic analysis of production. In *Production economics: A dual approach,* Vol. 1. ed. M. Fuss and D. McFadden. Amsterdam: North-Holland, pp. 217–268.

Gannon, C., and L. Zhi. 1997. *Poverty and transport.* Discussion Paper TWU-30. Washington, D.C.: World Bank.

Howe, J. D. 1981. *The impact of rural roads on poverty alleviation: A review of the literature.* World Employment Program Research Working Paper No. 106. Geneva: International Labor Organization.

———. 1997. *Transport for the poor or poor transport?: A general review of rural transport policy in developing countries with emphasis on low-income areas.* Geneva: International Labor Organization.

Howe, J., and P. Richards. 1984. *Rural roads and poverty alleviation.* London: Intermediate Technology.

Hussain, A. 2003. *Urban poverty in China: Measurement, patterns, and policies.* Geneva: International Labor Office.

Hussain, A., P. Lanjouw, and N. Stern. 1994. Income inequalities in China: Evidence from household survey data. *World Development* 22 (12): 1947–1957.

Jacoby, H. G. 2000. Access to markets and the benefits of rural roads. *The Economic Journal* 110: 713–737.

Kanbur, R., and X. Zhang. 1999. Which regional inequality? The evolution of rural-urban and inland-coastal inequality in China, 1983–1995. *Journal of Comparative Economics* 27: 686–701.

———. 2004. *Fifty years of regional inequality in China: A journey through central planning, reform and openness.* Research Paper 2004/50. Helsinki: United Nations University–World Institute for Development Economics Research.

Kessides, C. 1993. The contributions of infrastructure to economic development: A review of experience and policy implications. Mimeo. Washington, D.C.: World Bank.

Khan, A. R. 1997. Poverty in China in the period of globalization. Mimeo. University of California, Riverside.

Khan, A. R., and C. Riskin. 2001. *Inequality and poverty in China in the age of globalization.* New York: Oxford University Press.

———. 2004. China's household income and its distribution, 1995 and 2002. Mimeo. University of California, Riverside.

Khandker, S. 1989. *Improving rural wages in India.* Policy, Planning, and Research Working Paper 276. Washington, D.C.: World Bank.

Khandker, S., V. Lavy, and D. Filmer. 1994. *Schooling and cognitive achievements of children in Morocco.* World Bank Discussion Paper 264. Washington, D.C.: World Bank.

Kohli, U. 1982. A gross national product function and the derived demand for imports and supply of exports. *Canadian Journal of Economics* 18: 369–386.

Kwon, E. 2001. Infrastructure, growth, and poverty reduction in Indonesia: A cross-sectional analysis. Mimeo. Manila: Asian Development Bank.

Limao, N., and A. J. Venables. 1999. *Infrastructure, geographical disadvantage and transport costs.* Policy Research Working Paper Series 2257. Washington, D.C.: World Bank.

Lin, J. Y., F. Cai, and Z. Li. 1996. *The China miracle: Development strategy and economic reform.* Hong Kong Centre for Economic Research.

Lin, S., and S. Song. 2002. Urban economic growth in China: Theory and evidence. *Urban Studies* 39 (12): 2251–2266.

Liu, Z. 1999. *Guang Ming Daily,* April 7, 1999.

Mahapa, S. M., and M. Mashiri. 2001. Social exclusion and rural transport: gender aspects of a road improvement project in Tshitwe, Northern Province. *Development Southern Africa* 18(3): 365–376.

Malmberg, C. C., A. Ryan, and L. Pouliquen. 1997. *Rural infrastructure services for development and poverty reduction.* Washington, D.C.: World Bank. <http://www.worldbank.org/wb/sdruralpoverty/infrastructure/pdf/malmberg.pdf>.

Ministry of Agriculture. 1980–2003. *China agricultural yearbook.* Beijing: China Agricultural Publishing House.

———. 1995–2003. *China rural energy yearbook.* Beijing: China Agricultural Publishing House.

———. 1985–2003. *China agricultural development report.* Beijing: China Agricultural Publishing House.

Ministry of Education. 1984–2003. *China education yearbook.* Beijing: People's Education Press.

Ministry of Electric Power. 1990–2003. *China power yearbook.* Beijing: China Electric Power Publishing House.

Ministry of Transportation. 1984–2003. *China transportation yearbook.* Beijing: China Statistical Publishing House.

Ministry of Water Conservancy. 1991–2003. *China water conservancy yearbook.* Beijing: Water and Power Publishing House.

Minten, B., and S. Kyle. 1999. The effects of distance and road quality on food collection, marketing margins, and traders' wages: Evidence from former Zaire. *Journal of Development Economics* 60: 467–495.

Nagaraj, R., A. Varoudakis, and M. A. Veganzones. 2000. Long-run growth trends and convergence across Indian states. *Journal of International Development* 12: 45–70.

Ratner, J. B. 1983. Government capital and the production function for U.S. private output. *Economic Letters* 13: 213–217.

Ravallion, M., and S. Chen. 1997. *When economic reform is faster than statistical reform: Measuring and explaining inequality in rural China.* Washington, D.C.: World Bank, Policy Research Department.

Rawski, T. G. 2001. What is happening to China's GDP statistics? *China Economic Review* 12 (4): 347–354.

Riskin, C. 2004. *The fall in Chinese poverty: Issues of measurement, incidence and cause.* Presented at the Keith Griffin Festschrift Conference at Political Economy Research Institute, University of Massachusetts, Amherst, April 23–24, 2004.

Riverson, J., J. Gaviria, and S. Thriscutt. 1991. *Rural roads in Sub-Saharan Africa—Lessons from World Bank experience.* World Bank Technical Paper No. 141. Africa Technical Department Series. Washington, D.C.: World Bank.

Robinson, R. 2001. Decentralization of road administration: Case studies in Africa and Asia. *Public Administration and Development* 21 (1): 53–64.

Rozelle, S. 1994. Rural industrialization and increasing inequality: Emerging patterns in China's reforming economy. *Journal of Comparative Economics* 19 (3): 362–391.

Songco, J. A. 2002. *Do rural infrastructure investments benefit the poor? Evaluating linkages: A global view, a focus on Vietnam.* Policy Research Working Paper Series 2796. Washington, D.C.: World Bank.

State Statistical Bureau (SSB). various years. *China development report.* Beijing: China Statistical Publishing House.

———. 1982–2003. *China statistical yearbook.* Beijing: China Statistical Publishing House.

———. 1984–2003. *China rural statistical yearbook.* Beijing: China Statistical Publishing House.

———. 1985–99. *China agricultural development bank yearbook.* Beijing: China Statistical Publishing House.

———. 1990. *Historical statistical materials for provinces, autonomous regions and municipalities (1949–1989).* Beijing: China Statistical Publishing House.

———. 1997a. *The gross domestic product of China, 1952–95.* Dalin: Dongbei University of Finance and Economics Press.

———. 1997b. *Calculation methods of China's annual GDP.* Beijing: China Statistical Publishing House.

———. 2001. *China fixed asset investment statistical materials, 1950–2000.* Beijing: China Statistical Publishing House.

Stephan, A. 2000. Regional infrastructure policy and its impact on productivity: A comparison of Germany and France. *Konjunkturpolitik* 46 (940): 327–356.

Tsui, K. Y. 1991. China's regional inequality, 1952–1985. *Journal of Comparative Economics* 15 (1): 1–21.

Van de Walle, D. 2000. *Choosing rural roads investments to help reduce poverty.* Infrastructure Working Paper No. 2458. Washington, D.C.: Development Research Group, World Bank.

World Bank. 1994. *Infrastructure for development.* World Development Report 1994. New York: Oxford University Press.

———. 1999. *Transport in China—An evaluation of World Bank assistance.* Report No. 18865. Washington, D.C.: World Bank.

———. 2000. *China: Overcoming rural poverty.* Washington, D.C.: World Bank.

———. 2001, 2004. *World development report.* Washington, D.C.: World Bank.

Xian, Z., and L. Sheng. 2001. PRC's rural residents with consumption less than 860 yuan: Targeting group and characteristics. Memos, National Bureau of Statistics, Beijing.

Xinhua News Agency. 2003. Rural road construction speeded up. <www.china.org/English/2003/May/64719.htm>.

Zhang, X., and S. Fan. 2001. How productive is infrastructure? New approach and evidence from rural India. Environment and Production Technology Division. Discussion Paper No. 84. Washington, D.C.: International Food Policy Research Institute.

Zhou, T. 2001. Corruption in public investment and economic growth. *Reform of Economic System,* March/April.

Zhu, L. 1990. The impacts of public engineering in the economic growth of poor rural region, employment, and social services: A typical survey of the implementation of relieving people by project policy in 1980s. *Economic Research Journal* (in Chinese).

Made in the USA
Monee, IL
08 November 2026

67984628R00037